T0295417

# Seismic-Based Prediction Technologies for Shale Gas Sweet Spots

# Seismic-Based Prediction Technologies for Shale Gas Sweet Spots


Edited by

Qingcai Zeng

PetroChina Research Institute of Petroleum Exploration and Development, China

Sheng Chen

PetroChina Research Institute of Petroleum Exploration and Development, China

Majia Zheng

PetroChina Southwest Oil & Gas Field Company, China

Xiujiao Wang

PetroChina Research Institute of Petroleum Exploration and Development, China

Liwei Jiang

PetroChina Zhejiang Oilfield Company, China

Qiyong Gou

Shale Gas Research Institute of PetroChina Southwest Oil & Gasfield Company, China

*Published by*

World Scientific Publishing Co. Pte. Ltd.
5 Toh Tuck Link, Singapore 596224
*USA office:* 27 Warren Street, Suite 401-402, Hackensack, NJ 07601
*UK office:* 57 Shelton Street, Covent Garden, London WC2H 9HE

Library of Congress Control Number: 2023053707

**British Library Cataloguing-in-Publication Data**
A catalogue record for this book is available from the British Library.

Originally published in Chinese by Petroleum Industry Press


Seismic Prediction Technology of Shale Gas Sweet Spots


**SEISMIC-BASED PREDICTION TECHNOLOGIES FOR SHALE GAS SWEET SPOTS**



ISBN 978-981-12-8317-8 (hardcover)
ISBN 978-981-12-8318-5 (ebook for institutions)
ISBN 978-981-12-8319-2 (ebook for individuals)

For any available supplementary material, please visit
https://www.worldscientific.com/worldscibooks/10.1142/13590#t=suppl

Desk Editors: Balasubramanian Shanmugam/Amanda Yun

Typeset by Stallion Press
Email: enquiries@stallionpress.com

https://doi.org/10.1142/9789811283185_fmatter

# Foreword by Tianyue Hu

China enjoys enormous potential for shale gas resources, which are critical to the increase in the reserves and production of natural gas. To reduce the dependence on foreign oil and gas and ensure national energy security, it is vital at present and in the future for oil and gas researchers to achieve economies of scale in shale gas development. The seismic technologies for the prediction of shale gas sweet spots are the core technologies throughout integrated shale gas exploration and development. After years of practice and development, these technologies have formed their own system and unique characteristics, and have become an important discipline and development direction of oil and gas geophysical technology.

Zeng Qingcai and Chen Sheng, among others, have been engaged in the seismic prospecting of natural gas for more than a decade. They began to study the seismic technologies for the prediction of shale gas sweet spots at the very early stages of China's shale gas industry and participated in the construction of the Changning–Weiyuan national shale gas demonstration zone, gaining considerable practical experience and numerous results. Their primary innovative achievements using seismic data are as follows. They have developed the method and process for rock-physics modeling of shale gas reservoirs, establishing quantitative prediction templates of key parameters of shale gas reservoirs. They have improved the method for predicting pore pressure of shale gas reservoirs using seismic data, thereby increasing the prediction accuracy of overpressure shale gas sweet spots. They have developed the convolutional neural network (CNN)-based technique for the prediction of the gas content of shale reservoirs by introducing an artificial intelligence method. Furthermore, they have innovatively established the comprehensive quantitative prediction model of shale gas sweet spots using seismic data, realizing the quantitative seismic prediction of shale gas sweet spots.

This book, *Seismic-Based Prediction Technologies for Shale Gas Sweet Spots*, introduces relevant theoretical principles and technical methods of seismic technologies for the prediction of shale gas sweet spots comprehensively and systematically. It includes both the authors' latest research results and the latest development trends of these technologies at home and abroad, covering both theoretical research results and the application results in the production practice in main shale gas production areas in China. This book is expected to guide and promote future shale gas exploration and development, and contribute to the economies of scale in shale gas development in China.

https://doi.org/10.1142/9789811283185_fmatter

# Foreword by Yang Liu

The shale gas industry in China has rapidly developed, making China the third country to achieve economies of scale in shale gas development, after the United States and Canada. Despite its late start and weak foundation, China has achieved remarkable results in the shale gas industry by virtue of late-development advantages. As one of the obtained results, the seismic technology for the prediction of shale gas sweet spots has begun to play an important role and has become one of the core technologies for shale gas exploration and development in China.

Zeng Qingcai and Chen Sheng, among others, have long been engaged in the seismic prospecting of natural gas and are the first group of geophysicists in China specializing in the prediction of shale gas sweet spots using geophysical methods. They participated in the exploration and discovery of the Changning and Weiyuan shale gas fields and the construction of the Changning–Weiyuan national shale gas demonstration zone, gaining many innovative results. Most importantly, they have developed six innovative seismic technologies for shale gas reservoirs, including the rock-physics modeling of reservoirs, the quantitative prediction of total organic carbon (TOC) content, the prediction of pore pressure and overpressure shale gas sweet spots, the artificial intelligence (AI)-based prediction of gas content, and the comprehensive quantitative prediction of shale gas sweet spots. Furthermore, they have established a series of seismic technologies and processes for the comprehensive quantitative prediction of shale gas sweet spots. This book details the innovation of these technologies and their application results in China during recent years.

This book systematically introduces the latest development trends of seismic technologies for the prediction of shale gas sweet spots at home and abroad, and incorporates the research and development of new technologies and their application results in the production practice in primary shale gas production areas in China. This book processes innovative characteristics. For instance, cutting-edge technologies such as AI and machine learning have been applied to the prediction of shale gas sweet spots, achieving excellent results in shale gas production. Moreover, the seismic-geological-engineering integrated technological philosophy proposed in this book has been widely applied to shale gas production in China. This book will help to guide and promote the technological innovation and production practice of shale gas exploration and development.

https://doi.org/10.1142/9789811283185_fmatter

# About the Editors

**Qingcai Zeng** is the Director of the Geophysical Technology Department at the Research Institute of Petroleum Exploration and Development, PetroChina. His research interests include seismic data analysis, processing, and gas reservoir detection. He published more than 45 papers and 4 monographs. Besides, he holds 10 authorized invention patents and 4 software copyrights. He was awarded provincial and ministerial science and technology awards 6 times. Furthermore, he was awarded bureau-level science and technology awards 11 times.

**Sheng Chen** is the Vice Director of the Geophysical Technology Department at the Research Institute of Petroleum Exploration and Development, PetroChina Company Limited. His research interests include seismic data interpretation, reservoir prediction, gas reservoir detection, and unconventional seismic techniques. He published more than 38 papers and 4 monographs. He also holds 5 authorized invention patents and 2 software copyrights. He has been awarded provincial and ministerial science and technology awards 4 times and bureau-level awards 10 times.

**Majia Zheng** is the Deputy Chief Geologist at PetroChina Southwest Oil and Gasfield Company (SWOG). His work focuses on unconventional gas development. He has published more than 20 papers related to shale gas and tight gas. He has been awarded provincial and ministerial science and technology awards several times and a bureau-level award once.

**Xiujiao Wang** is a Senior Geophysicist of the Geophysical Technology Department at the Research Institute of Petroleum Exploration and Development, PetroChina. She specializes in unconventional oil and gas prediction. She has published more than 20 papers and 1 monograph. She also holds 2 authorized invention patents and 2 software copyrights. She has been awarded provincial-level awards twice and a bureau-level award once.

**Liwei Jiang** is a Senior Engineer at PetroChina Zhejiang Oilfield Company, specializing in comprehensive evaluations of unconventional oil and gas geology and seismic exploration. He has published more than 10 papers to date. He has been awarded one provincial and ministerial science and technology award.

**Qiyong Gou** is a Senior Engineer at Shale Gas Research Institute of PetroChina Southwest Oil and Gasfield Company, China. He focuses on unconventional oil and gas seismic interpretation and prediction, especially in shale gas of the Longmaxi Formation in the Sichuan Basin. He published eight scientific papers and holds one authorized invention patent. He was awarded bureau-level science and technology awards eight times.

https://doi.org/10.1142/9789811283185_fmatter

# Acknowledgments

The preparation team for this manuscript comprised the following:

**Leaders**: Zeng Qingcai, Chen Sheng, Zheng Majia

**Deputy leaders**: Wang Xiujiao, Jiang Liwei, Gou Qiyong

**Reviewers**: Dong Shitai, Gan Lideng

**Members**: Zeng Tongsheng, Yang Qing, He Pei, Wang Xing, Jiang Ren, Huang Jiaqiang, Bao Shihai, Dai Chunmeng, Yang Yadi, Li Xinyu, Li Wenke, Zhang Xiaowei, Song Yaying, Zhang Lianqun, Hu Xinhai, Li Xuan, Guo Wei

https://doi.org/10.1142/9789811283185_fmatter

# Contents

https://doi.org/10.1142/9789811283185_0001

# Chapter 1

# Significance and Current State of Seismic Technologies for the Prediction of Shale Gas Sweet Spots

China has considerable potential for shale gas resources, which are critical to the increase in the reserves and production of natural gas. Shale gas refers to the natural gas occurring within organic-rich shale horizon formations in the form of adsorbed and free gases, whose accumulation is free from the control of buoyancy. Shale gas sweet spots are areas with a moderate burial depth, high organic carbon content, high maturity, favorable preservation conditions, and high content of brittle minerals. In these areas, complex fracture networks can be formed through hydraulic fracturing. Shale gas sweet spots act as preferences for the effective development and high and stable production of shale gas. This chapter is composed of three sections. The first section introduces the distribution of shale gas resources and the significance of seismic technologies for the prediction of sweet spots. The second section states the current status and development trend of the seismic technologies for the prediction of sweet spots at home and abroad. Lastly, based on the exploration, development, and production of shale gas in China, the third section introduces the research contents and ideas of the seismic technology suitable for the prediction of shale gas sweet spots in southern Sichuan Province, China.

## 1.1. Significance of research into seismic technologies for the prediction of shale gas sweet spots

The production success of shale gas in North America has exerted a significant influence on the global natural gas supply. The United States has achieved domestic natural gas supply owing to the commercial development of shale gas. The U.S. shale gas revolution, which is the combination of hydraulic fracturing and horizontal drilling, has changed the global energy supply pattern. It enabled the U.S. to increase its shale gas production from $200 \times 10^8$ $m^3$ in 2003 to $4{,}447 \times 10^8$ $m^3$ in 2016, which accounted for over 50% of the total natural gas production. It contributed to the discovery of seven shale gas fields with an annual production exceeding $200 \times 10^8$ $m^3$ in the U.S. Moreover, the increased amplitude of 5.5% of the shale gas production in 2016 compared to 2015 remains and continues to increase, as predicted by relevant agencies (Zou, 2015). Presently, the United States has developed a series of key technologies with 3D seismic technologies for the prediction of sweet spots, high-volume hydraulic fracturing combined with horizontal drilling, and micro-seismic monitoring as cores.

Similar to the United States, China has considerable shale gas potential. According to the latest resource evaluation results, the recoverable resources of shale gas are up to $10 \times 10^{12}$–$25 \times 1012$ $m^3$ and the productivity of shale gas is predicted to reach $300 \times 10^8$ $m^3$ in 2020 in China. However, shale gas is more difficult to exploit in China. Furthermore, the exploration and development of shale gas in China started late and are still "young". Correspondingly, the technologies for the prediction of shale gas sweet spots in China are also in their early stages.

In China, the research into the technologies of non-conventional energy started late, and basic research, such as on the response mechanisms of shale reservoirs, is still lagging. Moreover, the technologies for the assessment and quantitative prediction of shale gas sweet spots in China are mostly developed by referring to foreign mature experience and software technologies. In some cases, the technologies for the prediction of conventional natural gas reservoirs are used in the shale gas field. Despite the progress made, they cannot meet the demand for shale gas production due to their poor application effects (Zou, 2016). Therefore, there is an urgent need to make breakthroughs.

Meanwhile, an in-depth exploration into these technologies is crucial for achieving economies of scale early in shale gas development in China. The sweet spots prediction is the most important in shale gas exploration and development. For instance, the geophysical technologies for sweet spots prediction act as a key technical factor that promotes breakthroughs in shale gas development in North America. The lateral and vertical distribution of sweet spots can be predicted mainly by using logging and seismic technologies, respectively. These technologies, which allow for the accurate prediction of the spatial distribution of shale gas sweet spots, can provide a basis for the layout and trajectory design of horizontal wells, thus improving benefits and reducing cost of shale gas development.

This book investigates the shale gas reservoirs of the Upper Ordovician Wufeng Formation and the Lower Silurian Longmaxi Formation (also referred to as the Wufeng–Longmaxi Formations) in southern Sichuan Province. Based on the logging and seismic responses of the shale gas reservoirs, it establishes a geophysical response model of shale gas sweet spots. It also studies the method for predicting key parameters of reservoirs and accordingly develops a seismic method for the quantitative prediction of shale gas sweet spots. Using this method, this book predicts the planar and spatial distribution of sweet spots in the reservoirs in the Wufeng–Longmaxi Formations. These researches are of great theoretical value for making breakthroughs in the method for predicting shale gas sweet spots in southern Sichuan Province. In addition, they will provide a scientific basis and technical support for achieving economies of scale early in shale gas development in southern Sichuan Province and even in the whole Sichuan Basin.

## 1.2. Current status of research into seismic technologies for the prediction of shale gas sweet spots

### 1.2.1. *Current status at home and abroad*

Many experts at home and abroad in this field have conducted massive research and exploration, achieving productive results in shale gas definition, fine-grained sediments, shale gas accumulation mechanisms, geological characteristics descriptions, shale gas enrichment laws, and the prediction of shale gas sweet spots.

With the advancement in shale gas exploration and development technologies, the definition of shale gas has become increasingly clear and accurate. In 2002, Curtis (2002) defined shale gas as "unconventional natural gas produced from the reservoir rock series dominated by organic-rich shales" at the annual meeting of the American Association of Petroleum Geologists (AAPG). In 2003, Chinese researcher Zhang *et al.* (2003) proposed a preliminary definition of shale gas for the first time in China, holding that shale gas can be rich in both argillaceous shales and siltstones. This definition was modified by Zhu (2007) in 2007, who considered that shale gas is primarily rich in argillaceous shales. In 2010, Zou (2010) described the characteristics and nanopores of shale gas reservoirs based on practical experience and redefined shale gas on this basis. In 2011, Zou (2011) further revised the definition of shale gas by adding the physical characteristics of shale gas reservoirs in response to actual production demand. In 2012, China National Petroleum Corporation (CNPC) put forward a new definition of shale gas after organizing many scholars to conduct relevant studies, discussions, and investigations. According to CNPC, shale gas refers to the natural gas occurring in organic-rich shale formations in the form of adsorbed and free gases, whose accumulation is free from the control of buoyancy. The accurate definition of shale gas is expressed more rigorously after excluding ambiguity. The precise understanding of shale gas allows for more accurate academic and technical exchanges. Meanwhile, a series of keywords used to describe the physical and lithologic characteristics of shale gas reservoirs have been proposed and normalized, greatly promoting the improvement of relevant basic theories and the production practice of shale gas.

As the exploration and development of shale gas deepen at home and abroad, fine-grained sedimentology has become a hot topic and frontier in the sedimentological study in China. Fine-grained sedimentology focuses on the material composition, structures, textures, classification, genesis, sedimentary processes, and distribution patterns of fine-grained sedimentary rocks (Magilligan, 1992). Fine-grained sediments are sedimentary rocks composed of fine-grained particles with a particle size of less than 0.0625 mm, as defined by the sedimentological community. Fine-grained sediments have complex components dominated by clay minerals, carbonate minerals, feldspars, felsic minerals, and other authigenic minerals. Fine-grained sedimentary rocks include common sediments such as shales, mudstones, siltstones, and clays. Based on the introduction and assimilation of foreign concepts, research methods, and technologies, Chinese researchers have conducted massive studies on fine-grained sediments in China's major sedimentary basins. As a result, they have obtained numerous innovative achievements and understandings of the sedimentary system, lithology, and lithofacies of fine-grained sedimentary rocks, as well as the physical characteristics of hydrocarbon reservoirs. Jiang *et al.* (2013) proposed detailed academic concepts, terms, and classification schemes of fine-grained sediments. Based on this, they summarized the characteristics of fine-grained sediments of siliceous carbonates from the perspective of sedimentary dynamics. Moreover, they preliminarily established the formation and distribution patterns of fine-grained sedimentary rocks in the Dongying Depression and identified the sequence framework of fine-grained sediments in this depression with the relative lake-level changes and the paleoclimatic and provenance characteristics at that time as constraints. Given the unique geological characteristics of unconventional oil and gas in China, Jia (2014) proposed the following suggestions for the geological study of unconventional oil and gas based on massive case studies. First, it is necessary to determine the high-resolution division scheme of the fine-grained sedimentary system. Then, the correlation between conventional and fine-grained sedimentary systems should be established to further determine the differences between the two systems. Yuan *et al.* (2013) conducted an in-depth study on the direction, latest advances, and development trends of the study on fine-grained sedimentary rocks. They researched the fine-grained sedimentary rocks in the Yanchang Formation in the Ordos Basin and summarized the distribution and spatial changes of fine-grained sedimentary systems in this formation. Furthermore, they determined the sedimentary patterns of lacustrine-facies organic-rich shales in this formation. Accordingly, they proposed that the main factors controlling organic-rich shales in the formation include paleowater depth, sedimentary facies belt, characteristics of lacustrine flow, and anoxic environment. Overall, fine-grained sedimentology booms mainly because the exploration of unconventional oil and gas (e.g., shale gas) has posed higher requirements for sedimentology. It is still at the earliest stage of rapid development, and related theories, definitions, and patterns are still being improved.

The genetic types of shale gas include thermal, biological, bio-thermal origins, or any combination thereof. Biogenetic gas is mainly sourced from bacterial microorganisms, which are primarily rich in atmospheric precipitation. Thermogenic gas involves three geneses: (1) the thermal degradation of kerogens when they are deeply buried or at a high temperature, (2) the secondary cracking of low-maturity biogas under high-temperature and high-pressure conditions, and (3) the secondary cracking of high-mature crude oil or asphalt. Shale gas has complex and variable accumulation processes, which represent a series of natural gas accumulation mechanisms to a certain extent and are significant for study. Owing to different accumulation conditions, the shale gas accumulation mechanisms include piston-type migration and accumulation, adsorption, and displacement migration and accumulation. Based on different accumulation mechanisms, the natural gas accumulation process can be divided into three stages. In the first stage, natural gas is formed and adsorbed under appropriate temperature and pressure conditions. In the second stage, natural gas is discharged from source rocks. In the third stage, natural gas migrates and is displaced in reservoirs. Shale gas primarily accumulates in the first two stages.

According to the summary of the successful experience in shale gas exploration and development in North America and China, shale gas reservoirs have unique geological characteristics compared with conventional natural gas reservoirs.

In terms of lithology, shale gas reservoirs mostly consist of dark or black asphaltenic or organic-rich shales (high carbon content). From the view of mass fractions, these shales are composed of about 15–25% of quartz mineral particles, 30–50% of clay minerals, and about 1–20% of organic matter. Generally, the main body of shale gas reservoirs is composed of dark mudstones, most of which have thin interlayers of light siltstones.

Shale gas primarily occurs in three ways in shales, namely, free gas, absorbed gas, and dissolved gas. Most shale gas is dominated by free gas, which mainly occurs in fractures and pores of shale reservoirs. The adsorbed gas is largely adsorbed in organic matter, clay minerals, and kerogen particles and on pore surfaces. The dissolved gas is relatively less in shale gas and mainly occurs in asphaltenes, kerogen, residual water, and crude oil. Nonetheless, shale gas may occur in other phases and states. The absorbed gas accounts for 20–85% of the total natural gas, significantly depending on basin types and tectonic movement history. This is apparently different from the occurrence state of conventional natural gas.

The accumulation of shale gas has its inherent characteristics, which are primarily reflected by the *in-situ* storage of the natural gas produced by mature source rocks. Furthermore, shale gas has complex accumulation mechanisms. During shale gas accumulation, all processes including dissolution, adsorption, piston-type propulsion, and displacement migration occur to different extents. Since shales serve as the source rocks, reservoirs, and cap rocks of shale gas, shale gas has typical self-generating and self-storage accumulation mode and *in-situ* characteristics. Meanwhile, the distribution area of shale gas reservoirs is equivalent to the area of effective source rocks.

The physical properties of shale gas reservoirs are also significantly different from those of conventional natural gas reservoirs. Shale gas reservoirs have extremely low porosity and permeability, with total porosity of generally less than 10% and the permeability largely affected by natural fractures. Therefore, industrial capacity can only be obtained by artificial hydraulic fracturing during shale gas development. Ultramicro-pores have developed in shale gas reservoirs. They are dominated by nanopores, which have a large specific surface area and serve as the storage spaces for large amounts of adsorbed gas. Moreover, shale gas reservoirs are distributed in large areas and have good continuity, generally displaying the characteristics of regional distribution.

In terms of pore pressure, most shale gas reservoirs show apparent high-pressure anomalies due to the accumulation of large amounts of shale gas. Certainly, the pressure system and abnormally high pressure will change subject to later geological processes such as vertical tectonic movement. As indicated by the exploration and development of shale gas in China over the past two years, all high-output shale gas reservoirs show noticeable overpressure anomalies. Therefore, formation overpressure is an essential condition for the high output of shale gas in China.

In terms of geological conditions, critical factors that control the gas content of shale gas reservoirs and the occurrence state of shale gas include effective reservoir thickness, total organic carbon

(TOC), permeability, total porosity, the development degree of natural fractures, paleostructure characteristics, and the characteristics and distribution of faults. These controlling factors jointly determine the commercial value and economies of scale of shale gas reservoirs.

Different opinions and understandings on the enrichment laws of shale gas have also been formed. Different from conventional natural gas, shale gas has complex enrichment laws and it is difficult to identify the main controlling factors of shale gas enrichment. Chen (2009) listed effective reservoir thickness, TOC, and reservoir porosity (including pores and natural fractures) as three major controlling factors in the enrichment of shale gas sweet spots in the exploration and development of shale gas in southern Sichuan.

Studies reveal that the degree of shale gas enrichment is directly proportional to the effective thickness and distribution area of shale reservoirs. Large and thick reservoirs serve as the prerequisites for the industrial capacity of shale gas and provide material bases and storage spaces for the formation of adequate natural gas. On the premise of the same TOC and gas-generating intensity, the gas content and scale of shale gas reservoirs increase with an increase in the thickness and distribution area of organic-rich shales. It is generally believed that a large-scale commercial shale gas field can be classified so only when organic-rich shales present there are over 30 m thick, that is, when the effective reservoir thickness is greater than the hydrocarbon expulsion thickness. A sedimentary system is the main factor that controls the thickness and distribution area of shale reservoirs. Organic-rich shales mainly develop and accumulate in continental lacustrine basins, marine basins, and slopes of marine basins, which are favorable areas for shale gas exploration and development.

The degree of shale gas enrichment increases with an increase in TOC. TOC is one of the most important parameters for the assessment of shale gas reservoirs. On the one hand, TOC provides a material basis for shale gas generation and enrichment, and directly determines the hydrocarbon generation intensity and scale of shale gas. On the other hand, abundant organic matter provides carriers for adsorbed gas, and the amount of adsorbed gas is directly determined by TOC. Meanwhile, numerous organic pores developing in organic matter provide large storage spaces for free gas. Therefore, TOC also plays an important role in controlling the porosity and free gas content of shale reservoirs. Under the same basic geological conditions and the same evolutionary stage and degree, TOC determines the hydrocarbon generation intensity of shale gas and the contents of adsorbed gas and free gas. Under these conditions, TOC has a linear positive correlation with these three items, and the total gas content increases with an increase in TOC. Isothermal adsorption curves are frequently employed to describe the gas content of shale gas reservoirs. Experimental studies show that the morphology of isothermal adsorption curves is primarily determined by the type and evolutionary degree of parent materials. As parent material types become more favorable, the gas content gradually increases. In addition, other important factors influencing the gas content include the content and composition of clay minerals, water saturation, and organic matter maturity. The effects of these influencing factors on the gas content also need to be further studied.

Another important controlling factor in shale gas enrichment is the development degree of pores and microfractures in shale gas reservoirs. The more the developed pores and natural fractures, the greater the degree of shale gas enrichment. Compared to conventional natural gas reservoirs, shale gas reservoirs have much poorer physical properties, including extremely low porosity and permeability and extremely strong heterogeneity. Moreover, all free gas in shale gas reservoirs occurs in matrix pores and reservoir fractures, including induced fractures generated during drilling. The induced fractures can connect existing natural fractures, thus providing more storage and seepage spaces for natural gas. As the development degree of these fractures increases, the output of shale gas also increases. In other words, with an increase in the development degree of shale gas reservoirs in which complex fracture networks can be formed through hydraulic fracturing, the output and productivity of shale gas will increase. The content of brittle minerals plays the most important role in determining whether complex fracture networks can be formed in shale gas reservoirs through hydraulic fracturing. As the content of brittle minerals increases, complex fracture networks are more liable to be generated through hydraulic fracturing, and hence the output of shale gas increases.

Quantitative prediction technologies of shale gas sweet spots are significant for the exploration and development of unconventional gas reservoirs. Given this, researchers at home and abroad have made numerous valuable explorations. Based on a large number of petrophysical analyses, core tests, and forward modeling, Goodway *et al.* (2006) systematically studied the relationships of elastic parameters of rocks in shale reservoirs with the physical properties, gas-bearing property, and anisotropy of shales. Moreover, they predicted the distribution of shale gas sweet spots using the results of physical properties, gas-bearing property, and anisotropy of shales. According to stress analysis and *in-situ* stress prediction using seismic data, Arcangelo *et al.* (2011) predicted the favorable areas of reservoir stress. Based on this, they predicted the planar distribution of shale gas sweet spots, which were taken as the basis for the deployment of horizontal wells. It should be noted that the predicted sweet spots were not exact but are favorable areas for hydraulic fracturing. Treadgold *et al.* (2011) predicted the shale gas sweet spots in the Eagle Ford Shale by combining seismic and geological research technologies. Specifically, they described the fracture system and stress state in the study area in detail using seismic technologies and taking structural and sedimentary systems as constraints. Based on this, they predicted the shale gas sweet spots, achieving good results. Roxana *et al.* (2012) predicted the lithologic probability distribution of shale gas reservoirs through seismic inversion, aiming to provide bases for the deployment of horizontal wells and the optimization of the hydraulic fracturing scheme. This provided a new idea for predicting shale gas sweet spots. Larry *et al.* (2013) estimated the key parameters of shale gas reservoir assessment using geophysical and geostatistical methods, aiming to offer references for well layout and trajectory design of horizontal wells during shale gas development. This also provided an effective means for predicting the distribution of sweet spots in mature shale gas blocks. Baishali *et al.* (2014) conducted an in-depth comparative analysis of seismic facies analysis techniques using different algorithms and analyzed the seismic facies of shale gas reservoirs. Based on this, they selected favorable seismic facies to predict the areas of shale gas sweet spots. This provided a relatively effective method for predicting the distribution of shale gas sweet spots. Peake *et al.* (2014) studied methods for predicting shale gas sweet spots by combining microseismic and ground seismic techniques and rock mineral analysis. This provided an ideal and complete technical process for the prediction of shale gas sweet spots and also provided important insights into the prediction of shale gas sweet spots. Tinnin *et al.* (2015) proposed combining multiple data and methods to predict shale gas sweet spots and successfully predicted sweet spots in the Eagle Ford Shale. Liu (2013) proposed combining the petrophysical and geophysical methods to predict shale gas sweet spots and deeply explored the petrophysical analysis and geophysical prediction methods. This will provide some references for the geophysical prediction of sweet spots. Xiang *et al.* (2016) analyzed the geophysical characteristics of shale gas reservoirs and put forward the geophysical prediction technology from the non-seismic perspective. Taking the exploration in the large-shale Jiaoshiba shale gas field in the Sichuan Basin as an example, Chen *et al.* (2014) explored the application effect of high-precision 3D seismic exploration technology in shale gas prediction, which provided a useful technical idea for shale gas exploration.

### 1.2.2. *Challenges of seismic technologies for the prediction of shale gas sweet spots*

The prediction of shale gas sweet spots primarily faces the following scientific challenges. (1) Shale gas reservoirs feature the development of ultramicropores and beddings, complex pore structure, and significant amounts of adsorbed gas. There is no suitable seismic method for petrophysical modeling, resulting in the lack of scientific basis for the quantitative seismic interpretation of shale gas sweet spots. (2) Shale gas reservoirs show complex seismic wave fields due to their strong heterogeneity and anisotropy. Accordingly, shale gas sweet spots do not display definite seismic responses and cannot be effectively predicted using existing means. (3) The shale gas enrichment in southern Sichuan Province is controlled by formation overpressure. However, current seismic methods for predicting formation pressure cannot meet the demand for shale gas development due to their low prediction accuracy. (4) Shale gas sweet spots are subject to many controlling factors, which are not fully considered in the comprehensive methods for predicting and assessing conventional reservoirs using seismic technologies. Therefore, they cannot be quantitatively predicted using these methods.

Owing to these scientific and technical challenges, the current prediction accuracy of shale gas sweet spots fails to meet the demand for economies of scale in shale gas development. Therefore, there is an urgent need to study new methods and technologies.

## 1.3. Ideas of seismic technologies for the prediction of shale gas sweet spots

### 1.3.1. *Contents of research into seismic technologies for the prediction of shale gas sweet spots*

This book studies the Wufeng–Longmaxi Formations in the Changning area in the southern Sichuan Basin, China. Through logging and seismic-geologic comprehensive interpretation, key parameter prediction, and the comprehensive prediction and assessment of shale gas sweet spots, this book researches the comprehensive quantitative prediction technology of shale gas sweet spots and applies it to shale gas production areas in the Changning area. The detailed research contents are as follows.

(1) **Research on basic geological characteristics:** This part mainly includes collecting, studying, and absorbing previous research results. Afterward, these results are updated using the latest exploration and development data. This part is designed to comprehend the overall basic geological characteristics of target strata, such as the tectonics, sedimentation, faults, strata, and stress. It is also to preliminarily determine the prospect and favorable areas of shale gas exploration and development.
(2) **Research on the geophysical responses of shale gas reservoirs in the Longmaxi Formation, including log and seismic responses:** This part is to determine the sensitive elastic parameters and longitudinal distribution of shale gas sweet spots (i.e., the concentration sections). Moreover, it aims to determine the seismic responses of shale gas sweet spots, including the petrophysical response and seismic profile responses, in order to provide bases and guidance for the seismic prediction of shale gas sweet spots. The research content of this part mainly includes the log interpretation of reservoirs, the analysis and modeling based on rock physics, the forward modeling of reservoirs, and the fine calibration of reservoirs.
(3) **Research into the technologies for the prediction of key assessment parameters of shale gas sweet spots, including key geological and engineering parameters:** This part

is designed to improve the prediction methods of key parameters such as TOC and formation pressure through technological innovation. It is also to enhance the prediction accuracy and obtain accurate spatial distribution of various assessment parameters using advanced pre-stack seismic inversion. This part serves as the base of the follow-up comprehensive quantitative prediction of sweet spots. The research content of this part mainly includes the prediction of the burial depths of structures, TOC, pore pressure of reservoirs, the thickness of high-quality reservoirs, reservoir brittleness, and fractures.

(4) **Research and application of comprehensive quantitative prediction method of shale gas sweet spots:** This part is to develop a comprehensive quantitative assessment model and system based on the above three parts and the full integration of geological, geophysical, engineering, and production data dominated by seismic data. Furthermore, it is to achieve the quantitative comprehensive prediction of shale gas sweet spots in the production areas in southern Sichuan Province using the established model and system. It is also to conduct the trajectory design, optimization, and the selection of optimal targets of horizontal wells.

### 1.3.2. *Ideas and technical route of the prediction of shale gas sweet spots using seismic data*

(1) **Ideas for the prediction of sweet spots using seismic data:** First, study the basic geological conditions (e.g., sedimentary environment, tectonic evolution history, and stratigraphic distribution) of the Longmaxi Formation—the main occurrence formation of shale gas in southern Sichuan Province, aiming to provide necessary constraints for the prediction of shale gas sweet spots. Second, conduct log interpretation, research on single-well geophysical responses, and rock physics analysis of shale gas reservoirs based on log data processing. The geophysical responses of shale gas sweet spots can be clarified through the processing and interpretation of log data. Meanwhile, rock physics analysis can be used to select the optimal geophysical parameters that are sensitive to key assessment parameters

(e.g., gas content) and to establish the quantitative relationships between the geophysical parameters and key assessment parameters. Afterward, establish the geological model of shale gas reservoirs based on the analytical results of rock physics. Then, analyze the seismic wave field characteristics of reservoirs with different gas contents and comprehensively study the geophysical responses of shale reservoirs through seismic wave field simulation and amplitude variation and off-set (AVO) forward modeling. The purpose is to provide a basis for the prediction of shale gas sweet spots and the selection of the prediction methods. Third, predict the spatial and planar distributions of key assessment parameters (e.g., TOC, gas content, reservoir brittleness, formation pressure, and formation stress) using high-precision seismic response methods based on single-well assessment results and the analytical results of rock physics. Lastly, establish a model for the prediction and assessment of shale gas sweet spots that integrates geological, geophysical, and engineering factors. Then conduct the comprehensive quantitative prediction of shale gas sweet spots, select optimal favorable targets, and design, trace, and optimize the trajectory of horizontal wells in the favorable targets.

(2) **Technical route of the prediction of shale gas sweet spots using seismic data:** The technical steps and route are formulated as follows based on the above prediction ideas.

(i) Research on the basic geological characteristics of the Wufeng–Longmaxi Formations. Obtain more detailed basic geological characteristics of the study area by combining the latest data of exploration wells and outcrops with previous research results, aiming to (a) obtain an overall understanding of the basic geological characteristics such as the tectonics and sedimentation of the Wufeng–Longmaxi Formations in southern Sichuan Province; (b) study the tectonic evolution history of the Sichuan Basin and southern Sichuan Province and preliminarily determine the tectonic characteristics in the sedimentary period of the Wufeng–Longmaxi Formations; (c) conduct an in-depth study of the current tectonic characteristics of reservoirs in the Wufeng–Longmaxi

Formations in the study area, including the burial depth of structures and detailed development characteristics of fractures; and (d) study the sedimentary characteristics of strata, that is, study the sedimentary environment of the Wufeng–Longmaxi Formations using previous research results combined with logging and test data, determine the horizons favorable for the development of shale reservoirs, and preliminarily determine favorable areas and horizons for shale gas exploration and development.

(ii) Research on the characteristics and longitudinal distribution pattern of shale gas sweet spots. Determine the log responses and longitudinal sweet spot sections of shale gas reservoirs by analyzing log data and studying and applying the method of determining single-well key parameters; establish a reasonable geological model of shale gas reservoirs based on basic log data and logging interpretation results, as well as the understanding of basic geological conditions of the study area; conduct the analysis and simulation based on rock physics using the geological model to determine the sensitive parameters of sweet spots through rock physics analysis and to further determine the longitudinal distribution pattern of shale gas sweet spots; determine the seismic responses of shale gas sweet spots using seismic forward modeling, aiming to guide the seismic interpretation and description of shale gas sweet spots.

(iii) Research on the methods for predicting the spatial distribution of key geological and engineering parameters of shale gas reservoirs. Research and develop new geophysical technologies and improve existing technologies for the prediction of key assessment parameters of shale gas sweet spots based on 3D seismic data in combination with the specific and particular conditions of the study area; improve the prediction accuracy of key assessment parameters of shale gas sweet spots, such as TOC, reservoir pore pressure, and brittleness index; accurately predict the planar distribution of various assessment parameters of sweet spots.

(iv) Research on the comprehensive quantitative technology for the prediction of shale gas sweet spots. Establish a quantitative prediction method for shale gas sweet spots using mathematical optimization methods based on the prediction results of single parameters, as well as the empirical parameters and qualitative criteria formed in the shale gas exploration and development; conduct comprehensive quantitative prediction of the planar distribution of sweet spots; apply the prediction results to actual production to verify the reliability of the research results and modify and optimize the technologies developed in the research for better applicability.

This book is organized surrounding the study of basic geological characteristics, the study of response mechanisms, study of technologies, and technology application. The technical route is shown in Figure 1.1.

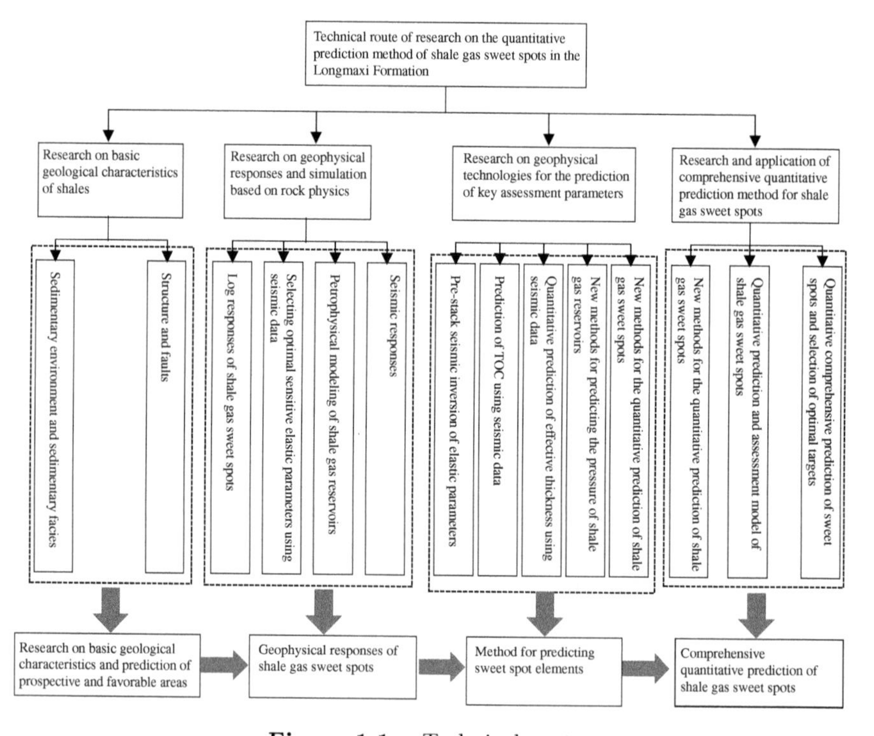


**Figure 1.1.** Technical route.

## Bibliography

Arcangelo S, Gabino C, Kevin C. Seismic reservoir characterization in resource shale plays: Stress analysis and sweet spot discrimination. *The Leading Edge*, 2011, 30(7): 758–764.

Beau T, McChesney MD, Bello H. Multi-source data integration: Eagle ford shale sweet spot mapping. *SEG Global Meeting Abstracts*, 2015, 624–631.

Chen GS, Dong DZ, Wang SQ, Wang YS. A preliminary study on accumulation mechanism and enrichment pattern of shale gas. *Natural Gas Industry*, 2009, 29(05): 17–21+134–135.

Chen ZQ. Quantitative seismic prediction technique of marine shale TOC and its application: A case from the Longmaxi Shale Play in the Jiaoshiba area, Sichuan Basin. *Natural Gas Industry*, 2014, 34(06): 24–29.

Chorn L, Yarus J, del Rosario-Davis S, Pitcher J. Identification of shale sweet spots using key property estimates from log analysis and geostatistics. *SEG Global Meeting Abstracts*, 2013: 1511–1523.

Curtis JB, Montgomery SL. Recoverable natural gas resource of the United States: Summary of recent estimates. *Aapg Bulletin*, 2002, 86 (10): 1671–1678.

Gading M, Wensaas L, Collins P. Methods for seismic sweet spot identification, characterization and prediction in shale plays. *Unconventional Resources Technology Conference*, USA (12 August to 14 August), 2013: 1402–1406.

Goodway B, Varsek J, Abaco C. Practical applications of P-wave AVO for unconventional gas Resource Plays Part 2. Detection of fracture prone zones with Azimuthal AVO and coherence discontinuity. Bibliogr, 2006.

Jia CZ, Zheng M, Zhang YF. Some key issues on the unconventional petroleum systems. *Acta Petrolei Sinica*, 2014, 35(01): 1–10.

Jiang ZX, Liang C, Wu J. Several issues in sedimentological studies on hydrocarbon-bearing fine-grained sedimentary rocks. *Acta Petrolei Sinica*, 2013, 34(6): 1031–1039.

Liu W, He ZH, Li KE, *et al.* Application and prospective of geophysics in shale gas development. *Coal Geology & Explora Tion*, 2013, 41 (6): 68–73.

Metzner D, Smith KL. Society of Exploration Geophysicists, American Association of Petroleum Geologists, Society of Petroleum Engineers. *Unconventional Resources Technology Conference*, Denver, Colorado, USA (12 August to 14 August), 2013: 256–262.

Peake N, Castillo G, Van de Coevering N, *et al.* Integrating surface seismic, microseismic, rock properties and mineralogy in the Haynesville Shale. *SEG Global Meeting Abstracts*, 2014, 343–353.

Roy B, Hart B, Mironova A, Zhou C, and Zimmer U. Integrated characterization of hydraulic fracture treatments in the Barnett Shale: The Stocker geophysical experiment. *Interpretation* 2014, 2(2): T111–T127.

Sahoo AK, Mukherjee D, Mukherjee A, *et al.* Reservoir characterization of Eagle Ford Shale through lithofacies analysis for identification of sweet spot and best landing point. *Unconventional Resources Technology Conference*, USA (12 August to 14 August), 2013: 1376–1383.

Treadgold G, *et al.* Eagle Ford shale prospecting with 3D seismic data within a tectonic and depositional system framework. *The Leading Edge*, 2011, 30(1): 2270.

Varga R, Lotti R, Pachos A, *et al.* Seismic inversion in the Barnett Shale successfully pinpoints sweet spots to optimize well-bore placement and reduce drilling risks. *SEG Technical Program Expanded Abstracts*, 2012: 1–5.

Xiang K, *et al.* Shale gas reservoir characterization and geophysical prediction. *Special Oil & Gas Reservoirs*, 2016, 23(02): 5–8+151.

Yuan GQ, Sun Y, Gao WD, *et al.* Development status of the shale gas geophysical prospecting technology. *Geology and Exploration*, 2013, 49 (05): 945–950.

Zhang JC, Xue H, Zhang DM, *et al.* Shale gas and its accumulation mechanism. *Geoscience*, 2003, 17 (4): 466–479.

Zhao WZ, Li JZ, Yang T, *et al.* Opportunities and challenges in the development of shale gas resources in China. In *Energy Forum organized by Chinese Academy of Engineering/National Energy Administration*, 2012.

Zhao WZ, Wang ZY, Wang HJ, *et al.* Shale gas: Changing the concept of oil and gas exploration and emerging a new clean gas resource. In *First Energy Forum organized by Chinese Academy of Engineering/National Energy Administration*, 2010.

Zou CN, Dong DZ, Yang H, Wang YM, Huang JL, Wang SF, Fu CX. Conditions of shale gas accumulation and exploration practices in China. *Natural Gas Industry*, 2011, 31 (12): 26–39. doi:10.3787/j.issn.1000-0976.2011.12.005.

Zou CN, Dong DZ., Wang SJ, *et al.* Geological characteristics and resource potential of shale gas in China. *Petroleum Exploration and Development*, 2010, 37(6): 641–653.

https://doi.org/10.1142/9789811283185_0002

# Chapter 2

# Geological Setting and Current Exploration and Development Status of Shale Gas Producing Areas in Southern Sichuan

Southern Sichuan, boasting the highest proven reserves and production of shale gas in China, is the only region in China that has achieved industrial production of shale gas. In this region, blocks such as Changning have taken the lead in shale gas development. Compared with the geological conditions of areas producing shale gas in North America, southern Sichuan has more complex surface and subsurface structures due to multi-phase tectonic movements, posing great challenges to the industrial development of shale gas. This chapter first introduces the tectonic evolution and the tectonic and sedimentary characteristics of different eras in southern Sichuan. Then, taking the Changning block as an example, this chapter focuses on the exploration and development history of shale gas in southern Sichuan, the difficulties and challenges in the seismic prediction of shale gas sweet spots, and corresponding technical countermeasures.

## 2.1. Basic geological characteristics of southern Sichuan

The Sichuan Basin is rhombic in shape. In terms of geotectonic location, it lies in the northwest of the Yangtze Platform and is a secondary tectonic unit of the platform. Its sedimentary strata include Paleozoic marine strata, Mesozoic and Cenozoic continental strata, and pre-Sinian metamorphic rocks as the basement. The Sichuan Basin took shape during the Indosinian, and its present tectonic features were formed after a series of Himalayan tectonic movements. It is one of the four major petroliferous bearing basins in China, covering an area of $18 \times 10^4$ km$^2$. Among all basins in China, the Sichuan Basin has the maximum cumulative production and the maximum proven reserves of conventional natural gas, as well as the highest proven reserves and production of shale gas.

### 2.1.1. *Tectonic characteristics*

Regarding tectonic classification, the Sichuan Basin is a typical superimposed basin according to previous exploration and development of conventional oil and gas in this basin. This basin was influenced by both the Tethyan and the Pacific tectonic domains, underwent the evolutionary stages of a craton and a foreland basin, and became a composite foreland basin formed by the superimposition of the craton basin onto the foreland basin in the Yangtze Platform during the Meso–Cenozoic.

Regarding tectonic cycles, the Sichuan Basin underwent multiphase tectonic movements during its formation and was frequently uplifted and subsided over millions of years of the Earth's history. The tectonic cyclicity shows that the tectonic movements in this basin can be divided into six tectonic cycles, including Yangtzeian, Caledonian, Hercynian, Indosinian, Yanshanian, and Himalayan cycles.

(1) **Yangtzeian cycle**: This tectonic cycle was formed at approximately 740.99–701.54 Ma and mainly includes the tectonic movements before the Sinian. A uniform basin basement was formed during this period, and the tectonic movements of this cycle mainly include the sedimentary folding inversion, the extrusion and intrusion of deep magmas, and the metamorphism and consolidation of sedimentary strata during the pre-Sinian. The Chengjiang tectonic movement, which is a representative movement in this cycle, spanned the middle and late stages of the Early Sinian and is marked by the intrusive rocks in the

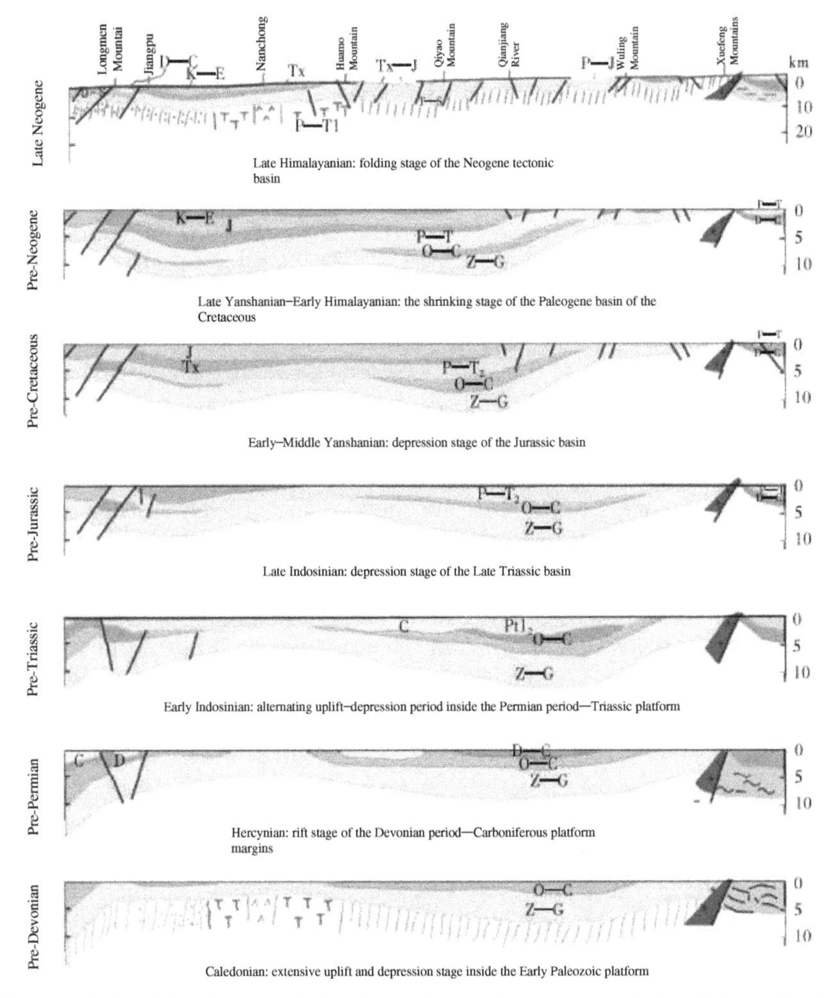


**Figure 2.1.** Sketch map of the formation and evolution of the Sichuan Basin (modified after Petro China Southwest Oil and Gasfield Company, 2011).

underlying strata of the Upper Sinian Dengying Formation and the purplish-red dacitic porphyry composed of extrusive rocks (Figure 2.1).

(2) **Caledonian cycle**: This tectonic cycle is divided into two stages, namely the Tongwan and the Late Caledonian tectonic movements. The former caused the upper strata of the Dengying Formation in the Sichuan Basin to be greatly uplifted and

denuded, leading to the disconformity between the Cambrian and the Sinian strata. It mainly occurred from the Late Sinian to the Early Cambrian. The Caledonian tectonic movement mainly caused the formation of the Leshan–Longnvsi paleo-uplift. During this stage, the Lower Paleozoic strata underwent large-scale folding, giving rise to the formation of a series of deep faults. A large uplift–depression pattern was further formed under the control of these deep faults. This tectonic cycle includes all the tectonic movements from the Late Sinian to the Late Silurian (Figure 2.1).

(3) **Hercynian cycle**: This tectonic cycle mainly features the continuous tectonic uplift, leading to the formation of multiple disconformities. Moreover, the early continuous tectonic uplift caused a large-area loss of the Devonian and the Carboniferous strata almost throughout the Sichuan Basin. The Dongwu tectonic movement occurred in the late stage, leading to a second large-scale large-amplitude tectonic uplift, which further resulted in the regional disconformities between the Upper Permian strata and the Lower Permian in the Sichuan Basin. This tectonic cycle has distinct lithologic indicators mainly including basalts and diabases, which are mainly distributed in the Huaying Mountain, northeastern Sichuan Province, and the Longquan Mountain (Figure 2.1).

(4) **Indosinian cycle**: This tectonic cycle can also be divided into two stages, namely, the Early Indosinian and the Late Indosinian tectonic stages, spanning the whole Triassic. During this tectonic cycle, transpressive stress predominates the whole Sichuan Basin. In the former stage, the large-scale transgression throughout the Sichuan Basin gradually ended, and large quantities of regressive deposits appeared. As a result, the basin was uplifted in a large range and was present as a large uplift–depression pattern consisting of alternate uplifts and depressions mainly in northeast trending. The Luzhou–Kaijiang paleo-uplift was formed along the Huayingshan fault zone during this stage. In the Late Indosinian tectonic stage, the tectonic movements in the Sichuan Basin were dominated by tectonic uplift accompanied by strong folding and faulting. The most important product of this stage was the Longmenshan tectonic belt. Another important

feature of the Indosinian cycle is that the Sichuan Basin gradually shifted from marine deposition toward continental deposition (Figure 2.1).

(5) **Yanshanian cycle**: This tectonic cycle spanned the whole Jurassic and the Cretaceous, during which continental sedimentary strata in the Sichuan Basin were mainly deposited. After the previous Indosinian cycle, the Sichuan Basin was uplifted on a large scale again and underwent folding due to the compression by its surroundings during the Yanshanian. Moreover, the large-scale tectonic uplift directly caused the denudation of the Late Jurassic strata, many of which were consequently in disconformable contact with Late Cretaceous strata. In the middle stage of the Yanshanian tectonic movements, the southeastern part of the basin strongly compressed toward the inside of the basin, directly causing the large-amplitude uplift of the piedmont of the Longmen Mountain (Figure 2.1).

(6) **Himalayan cycle**: This tectonic cycle mainly occurred during the Cenozoic and can be divided into two stages, namely, before and after the Neogene. The tectonic activities before the Neogene have significant effects on the present landforms of the Sichuan Basin. They caused the folding of all the strata deposited from the Sinian to the Paleogene, and the tectonic framework of the whole basin was roughly formed (Figure 2.1). The second tectonic movement of the Himalayan cycle mainly occurred after the deposition of the Neogene strata. Owing to the good tectonic inheritance, the structural features formed early were further strengthened, except for local adjustments. The present landforms of the Sichuan Basin were formed at this stage. Furthermore, compared with other tectonic movements, the Himalayan tectonic movements have the most profound effects on the formation of shale gas and conventional natural gas reservoirs in the Sichuan Basin (Figure 2.1).

The Sichuan Basin is surrounded by orogenic belts such as the Dalou, Daba, Micang, and Longmen mountains and can be divided into six tectonic units, namely, the high-steep fault-fold zone in eastern Sichuan, the low-gentle zone in northern Sichuan, the low-steep zone in western Sichuan, the gentle zone of the middle slope in

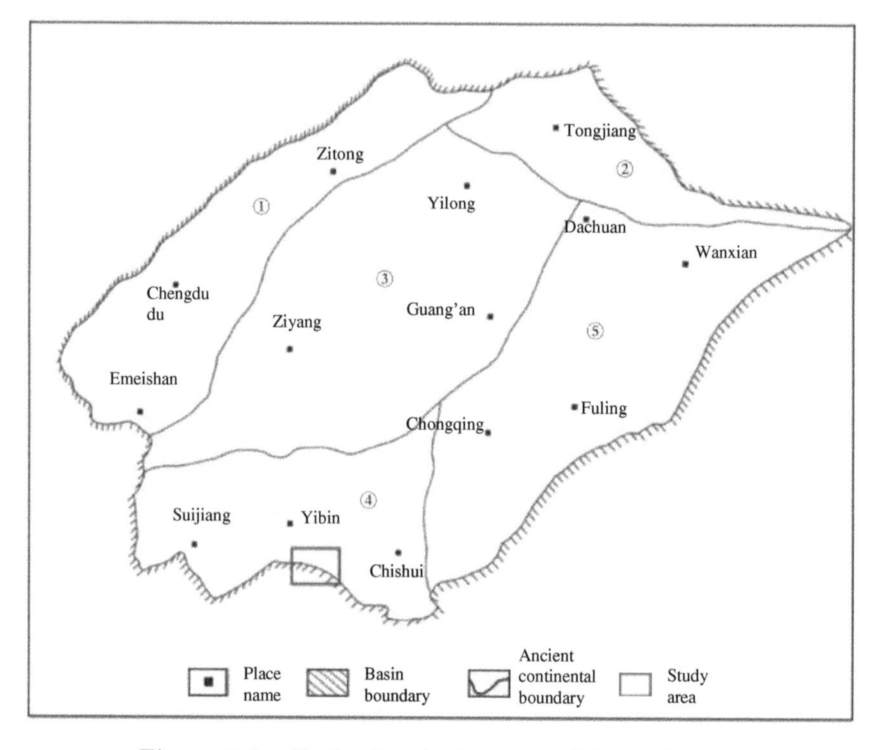


**Figure 2.2.** Regional tectonic setting of the study area.

the paleo-uplift in central Sichuan, the low-steep dome-shaped zone in southern Sichuan, and the high-steep fault-fold zone in eastern Sichuan.

The Changning shale gas field is located in the low-steep deformation zone of the middle uplift in the paleo-depression in southern Sichuan and the Loushan fold zone, as shown in Figure 2.2. The Changning anticline has mainly developed in the Changning shale gas field. It has a simple structure and an northwest–southeast strike overall, with a gentle SW wing and a steep NE wing. Affected by multi-phase tectonic 4.9 movements, the Changning shale gas field has mainly developed NE–SW- and nearly east-west-trending fault systems. Both of the fault systems consist of reverse faults, which are mainly on medium–small scales. Most of the faults disappear within the Silurian strata. Moreover, the bottom of the Longmaxi Formation has a depth of 1,250–2,550 m, as shown in Figure 2.3.

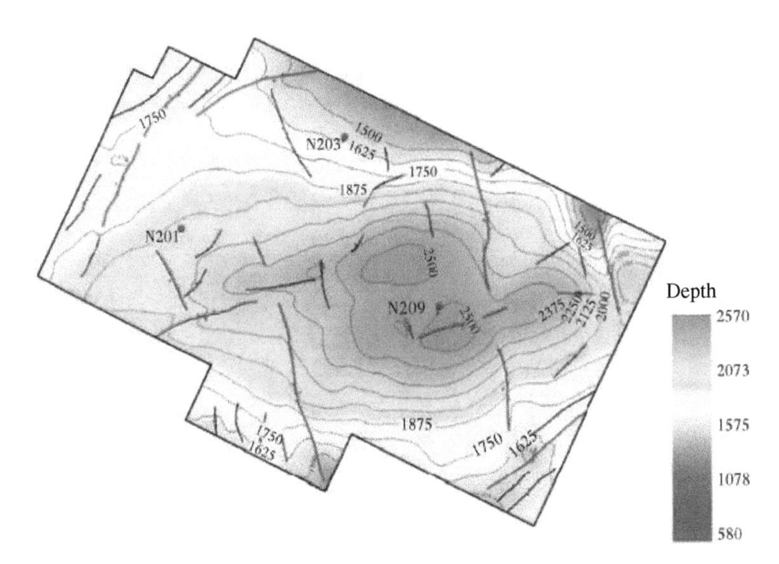


**Figure 2.3.** Tectonic map of the bottom of the Longmaxi Formation in the study area.

### 2.1.2. *Characteristics of sedimentary strata*

The Sichuan Basin is a typical superimposed basin in western China and contains oil and gas in multiple strata. Both marine and continental sedimentary strata have developed in the Sichuan Basin, with the marine sedimentary strata having an especially long depositional history. This basin consists of an extremely thick pre-Sinian basement, an extremely thick Sinian–Middle Triassic marine carbonate strata, and Late Triassic–Cretaceous continental clastic sediments. Four sets of high-quality marine source rocks developed from the Sinian to the marine sedimentary period of the Late Triassic. The present shale gas exploration in the Sichuan Basin focuses on the high-quality marine shale gas reservoirs of the Lower Silurian Longmaxi Formation and the Lower Cambrian Qiongzhusi Formation (Figure 2.4). Black shales in the two formations have mainly developed at their bottom. They are high-quality gas sources for conventional natural gas reservoirs of the underlying Sinian Dengying Formation and the overlying Carboniferous and Permian strata. Moreover, they can form high-quality shale gas reservoirs themselves.

The Cambrian Qiongzhusi Formation is widely distributed in the south of the Sichuan Basin. Its underlying Sinian Dengying

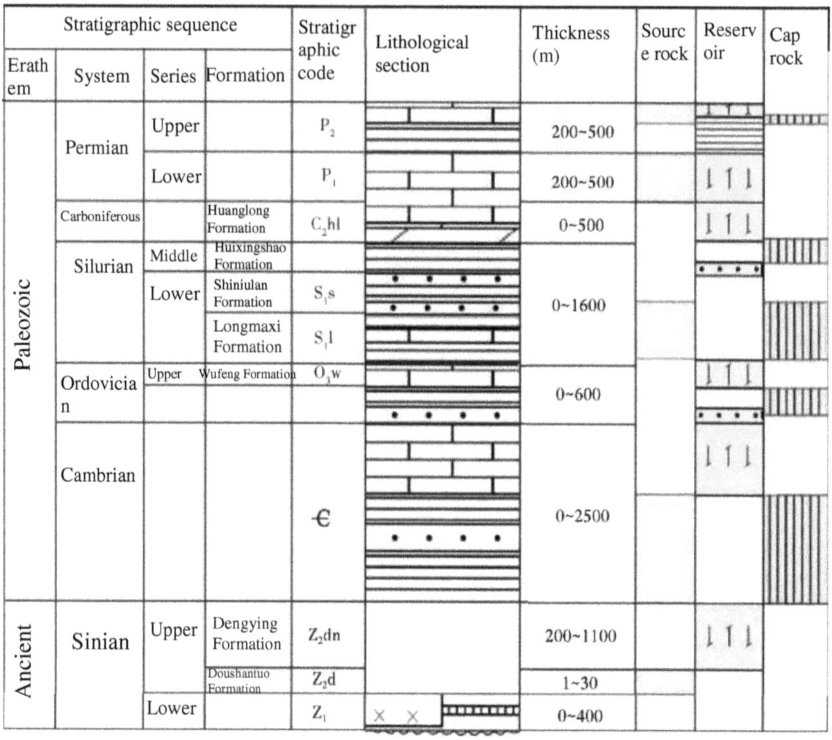

| Stratigraphic sequence | | | | Stratigraphic code | Lithological section | Thickness (m) | Source rock | Reservoir | Cap rock |
|---|---|---|---|---|---|---|---|---|---|
| Erathem | System | Series | Formation | | | | | | |
| Paleozoic | Permian | Upper | | $P_2$ | | 200~500 | | | |
| | | Lower | | $P_1$ | | 200~500 | | | |
| | Carboniferous | | Huanglong Formation | $C_2hl$ | | 0~500 | | | |
| | Silurian | Middle | Huixingshao Formation | | | 0~1600 | | | |
| | | Lower | Shiniulan Formation | $S_1s$ | | | | | |
| | | | Longmaxi Formation | $S_1l$ | | | | | |
| | Ordovician | Upper | Wufeng Formation | $O_3w$ | | 0~600 | | | |
| | Cambrian | | | € | | 0~2500 | | | |
| Ancient | Sinian | Upper | Dengying Formation | $Z_2dn$ | | 200~1100 | | | |
| | | | Doushantuo Formation | $Z_2d$ | | 1~30 | | | |
| | | Lower | | $Z_1$ | | 0~400 | | | |

**Figure 2.4.** The distribution and reservoir-cap rock assemblages of Lower Paleozoic strata in the study area.

Formation is the hot topic of the present exploration, and its overlying strata are Lower Ordovician. Cambrian strata can be divided into three parts under normal conditions, namely, the Lower Cambrian strata including the Qiongzhusi, Canglangpu, and Longwangmiao formations, the Middle Cambrian strata dominated by the Gaotai Formation, and the Middle–Upper Cambrian strata, which mainly include the Xixiangchi Group. When the Lower Cambrian Qiongzhusi Formation was deposited, the sea level was high, and the whole basin was in the period of maximum transgression. As a result, black and gray black shales were formed in the shelf environment, and they generally contain carbonate and siliceous interlayers.

The middle and upper parts of the Silurian strata in most areas of the Sichuan Basin have been denudated, with only the Lower Silurian

Longmaxi, Shiniulan, Xiaoheba, and Hanjiadian formations remaining. The Silurian strata are the main source rocks of their adjacent strata throughout the Upper Yangtze region. The black shales of the Lower Silurian Longmaxi Formation combined with the limestones of the Shiniulan Formation or the sandstones of the Xiaoheba Formation combined with the shales of the Hanjiadian Formation have constituted secondary source-reservoir-cap rock assemblages. Large quantities of graptolites have developed in the lower part of the Lower Silurian Longmaxi Formation. Moreover, the shales are generally black and have a high organic content. These phenomena indicate a highly reducing environment and deep-water shelf sedimentary facies overall at that time.

The Changning area has normal stratigraphic sequences in general. The outcrops in this area mainly include the Permian and the Triassic strata and the outcrops in the core of the Changning anticline in the study area include the Silurian, Ordovician, and Cambrian strata. The strata in this area include the Jurassic strata, the Triassic Xujiahe, Leikoupo, Jialingjiang, and Feixianguan formations, the Permian Changxing, Longtan, Maokou, Qixia, and Liangshan formations, the Silurian Hanjiadian, Shiniulan, and Longmaxi Formations, the Ordovician Wufeng, Linxiang, Baota, Dachengsi, and Luohanpo formations, the Cambrian Xixiangchi, Canglangpu, and Qiongzhusi formations, and the Sinian strata from surface to the basement, while Carboniferous and Devonian strata are lacking. Among them, the Wufeng and Longmaxi Formations were continuously deposited and are the target strata for present shale gas development.

This book focuses on the organic-rich shale reservoirs of the Longmaxi Formation at the bottom of the Lower Silurian strata. The black shales of the Longmaxi Formation are not only high-quality source rocks but also shale gas reservoirs. Moreover, they, together with their overlying tight limestone strata of the Shiniulan Formation, form a high-quality reservoir-cap rock assemblage, as shown in Figure 2.4. The research results of paleontological fossils show that a large number of graptolites occur throughout the Longmaxi Formation, indicating a highly reducing environment. The study area was in a deep-water shelf facies when the Longmaxi Formation was deposited, which is conducive to the formation of high-quality shales. The Longmaxi Formation in the study area has a stable thickness of approximately 500 m. It can be divided into two members, namely,

Long-I and Long-II members from bottom to top. The Long-I Member is a shale reservoir and has total organic carbon (TOC) of generally greater than 1%, while the Long-II Member has TOC of generally less than 1%. The Long-I Member can be subdivided into upper and lower submembers. Among them, the upper submember comprises gray shales with TOC of generally less than 2%. In comparison, the lower submember is mainly composed of black shales and carbonaceous shales, with developed bedding and TOC of generally greater than 2%. Therefore, the Longmaxi Formation is the most favorable horizon for shale gas exploration and development in the Sichuan Basin and is generally defined as a high-quality shale reservoir horizon. Figure 2.5 shows the lithology and TOC of the Longmaxi Formation and its top and bottom. The TOC is present as a square wave to make the TOC distribution more intuitive. Figure 2.5 shows that TOC tends to increase with the depth.

Structures were mainly controlled by the Guangxi Movement when the Wufeng Formation was deposited. During this period, the collision and merging of the Cathaysia Plate and the Yangtze Block eased, and a paleogeographical pattern of seven uplifts (the Motianling, Hanzhong, Central Sichuan, West Sichuan–Central Yunnan, Central Guizhou, Yunnan–Guizhou–Guangxi, and Xuefeng ancient continental blocks) surrounding one depression formed in the Sichuan Basin and its adjacent areas. In the early stage of the Longmaxi Formation (Rhuddanian–Early Aeronian; that is, the sedimentary period of the lower submember of the Long-I Formation), the sea level rapidly rose due to the melting of the Antarctic ice sheet and, accordingly, the entire southern Sichuan was in a sedimentary environment of a deep-water shelf with a large area of hypoxia. During the middle–late sedimentary period of the Longmaxi Formation (Middle Aeronian–Telychian; that is, the sedimentary period of the upper submember of the Long-I Member and the Long-II Member), violent collision and merging occurred between the Yangtze Plate and its surrounding plates. As a result, the subsidence and deposition center of the whole Sichuan Basin moved toward the central and northern parts of the basin, and the sea level dropped significantly. During this period, the Changning shale gas field transitioned from a deep-water shelf into a calcareous shallow-water shelf. Therefore, the sedimentary environment of the Wufeng–Longmaxi Formations can be divided into shallow- and deep-water shelf facies based on the

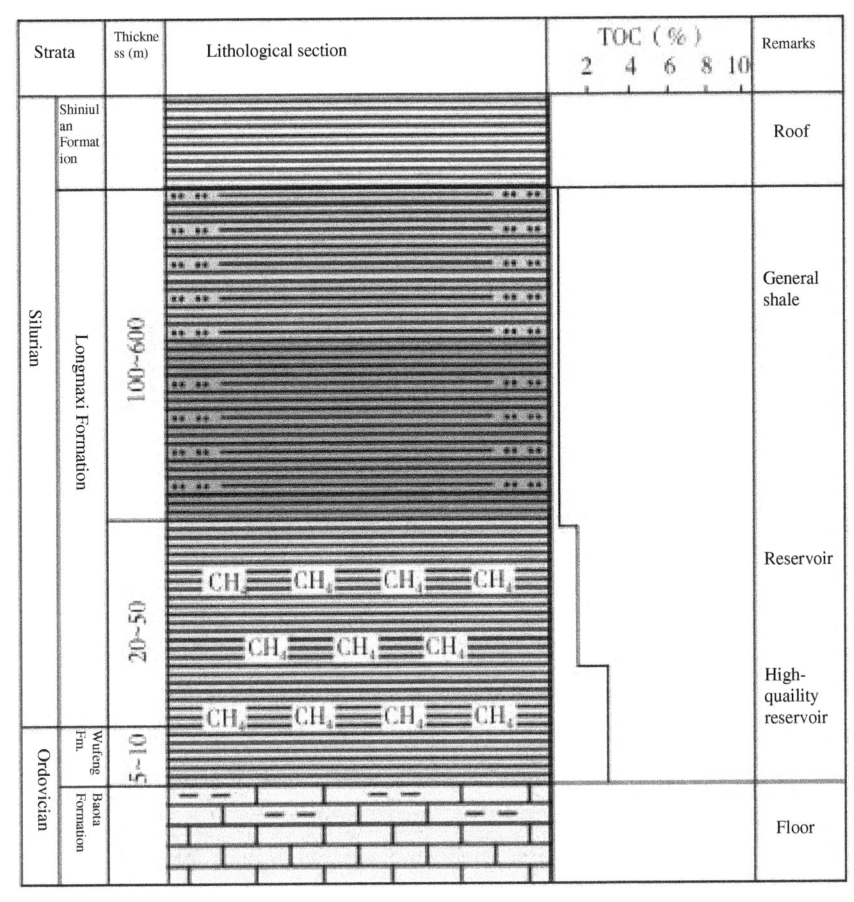


**Figure 2.5.** Lithological profile of the Longmaxi Formation as the target strata. *Note*: Fm: Formation.

characteristics of the formation, such as the hydrodynamic conditions during the sedimentary period of the formation, rock types and colors, lithological associations, sedimentary structures, original sedimentary environment, paleontological associations, and the content and composition of facies minerals (Figure 2.6).

The Long-I Member of the Wufeng Formation in the Changning shale gas field was mainly in a sedimentary environment of deep-water shelf facies during its formation. It is generally considered that this member was below the normal wave base in the deep-water shelf facies, with a water depth of generally less than or approximately

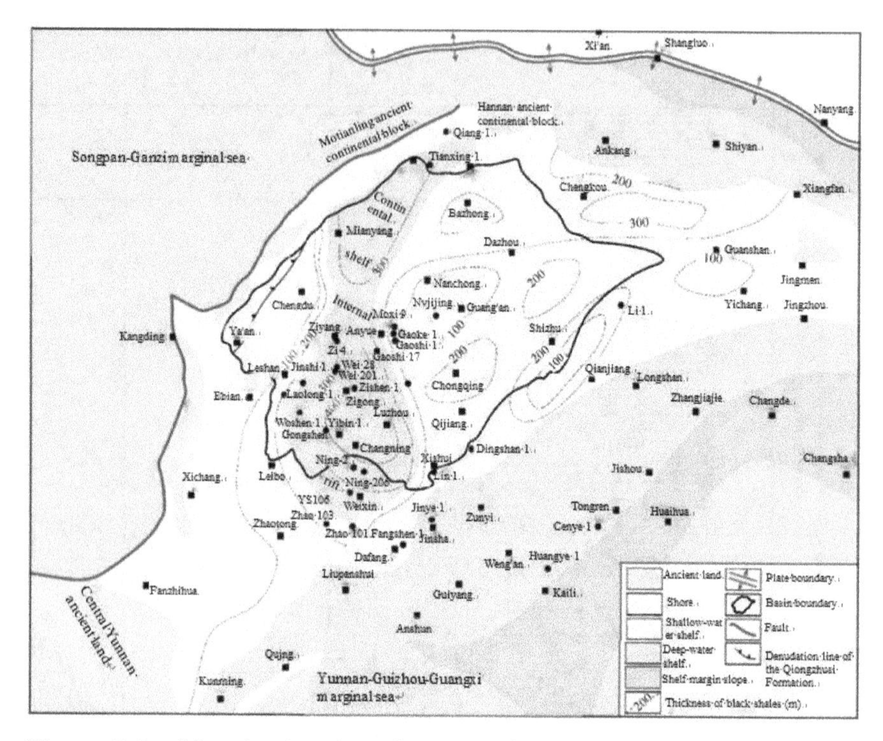


**Figure 2.6.** Map showing the sedimentary facies of the Longmaxi Formation in the Sichuan Basin and its adjacent areas.

200 m. In terms of planar distribution, this member was mostly located in or close to a coastal facies zone, and the shales deposited in this period are mainly characterized by dark clay-size clastic materials. The sedimentary facies of this member can be further divided into three types of microfacies, such as the organic-rich siliceous mud shelf (Table 2.1). Among them, the organic-rich siliceous mud shelf and organic-rich silty mud shelf are the most favorable microfacies for shale reservoir development.

The Longmaxi Formation in the Changning shale gas field generally has a deep-water shelf facies and its lithology is dominated by argillaceous shale. According to the drilling and seismic data, the thickness of the Longmaxi Formation is relatively stable in the shale gas field, with a residual thickness of 0–450 m generally. However, the Longmaxi Formation is denuded and exposed in the core of the Changning anticline, and its thickness gradually increases to the

**Table 2.1.** Sedimentary facies classification of the Long-I Member of the Wufeng Formation in Changning shale gas field.

| Sedimentary facies | Subfacies | Microfacies |
|---|---|---|
| Continental shelf | Shallow-water shelf | Argillaceous silty shelf |
| | | Calcareous silty mud shelf |
| | | Lime-mud silty shelf |
| | | Shallow-water silty mud shelf |
| | Deep-water shelf | Deep-water silty mud shelf |
| | | Organic-rich silty mud shelf |
| | | Organic-rich siliceous mud shelf |

north of the anticline. Moreover, the Longmaxi Formation in core shale gas-producing areas, such as areas of wells Ning-201 and Ning-209, has a thickness of approximately 300 m (Figure 2.7).

Based on the characteristics of sedimentary cycles, as mentioned earlier, the Longmaxi Formation can be divided into two members, namely, the Long-I and Long-II members, with a rhythmic interbed consisting of gray silts and shales at the bottom of the Long-II Member as the boundary. The overlying Long-II Member consists of grayish-black shales, while the underlying Long-I Member comprises black shales. The sedimentary cycles of the two members greatly differ and have a clear boundary. As shown in the natural gamma-ray (GR) log curve of the two members, the curve is in the shape of a bell with continuously decreased values for the top of the Long-I Member. For the part upward from the boundary between the two members, the gamma-ray values increase suddenly and then continuously decrease linearly and show a bell shape again.

During the sedimentary period of the Long-I Member, the sea level rapidly rose overall due to the melting of the Antarctic ice sheet. As a result, the entire southern Sichuan, including the study area, was in the sedimentary environment of a deep-water shelf with a large area of hypoxia. During the sedimentary period of the Long-II Member, with the subsidence center migrated toward the central and northern Sichuan and the seawater become shallow as a whole, the Changning shale gas field transitioned from a deep-water shelf to a calcareous shallow-water shelf.

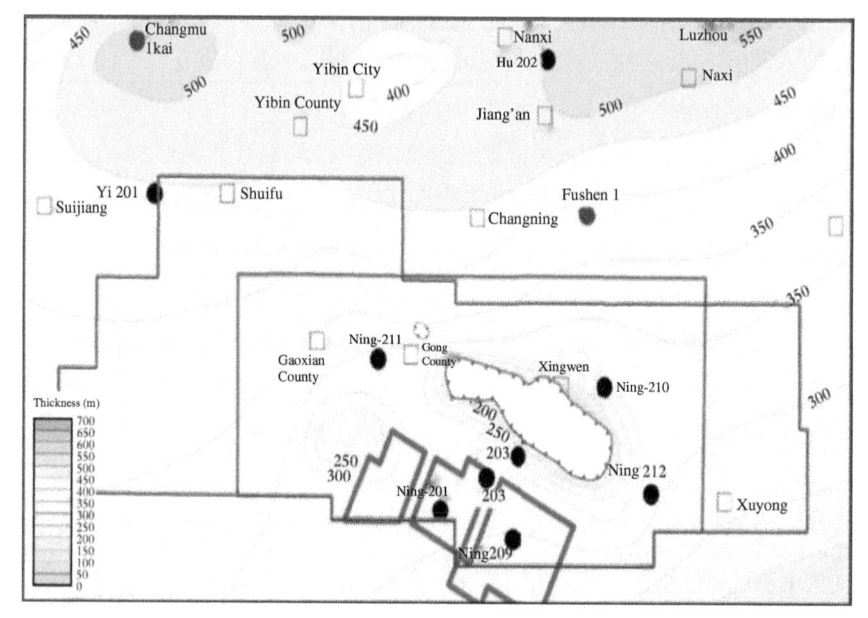


**Figure 2.7.** Map showing the residual thickness of the Silurian Longmaxi Formation in the Changning shale gas field.

The Long-I Member is a progressive reversed cycle with continuous regression. It can be divided into the upper and lower submembers based on the secondary cycle and lithological characteristics. The upper submember is a sedimentary cycle with gradually regressive highstand system tracts. This member contains a large section of interbeds consisting of sandy and argillaceous materials and silty interlayers, as well as a small number of graptolites. For the upper submember, a large section of interbeds or interlayers consisting of sandy and argillaceous materials have developed at the top with dark-gray shales at the bottom. The dark-gray shales at the bottom have a thickness of 105–200 m and serve as the boundary with the grayish-black shale in the underlying lower submember. The lower submember is a set of organic-rich black carbonaceous shales with well-developed laminations. It has large quantities of graptolite fossils of various shapes as well as pyrite nodules and strips, with a thickness of 16–48 m.

## 2.2. Exploration and exploitation: Current status of shale gas in the study area

### 2.2.1. *Exploration and exploitation history*

The shale gas exploration and exploitation in the Changning shale gas field have roughly experienced four stages, namely, resource assessment and exploration area selection, pilot tests of shale gas exploration and exploitation, the construction of shale gas demonstration areas, and industrial production.

(1) **Resource assessment and exploration area selection (2006–2009):** Petro China Southwest Oil and Gasfield Company (the Company) took the lead in conducting resource assessment and exploration area selection in the Sichuan Basin in 2006. Studies show that many sets of organic-rich black shales have developed in the Sichuan Basin. The Lower Paleozoic strata in the basin are favorable horizons, where shales are thick and are stably and widely distributed. Moreover, the quality of these shales is comparable with that of the typical shale gas in North America. Large numbers of gas shows were discovered during drilling, indicating considerable potential for shale gas exploration and exploitation. However, due to the unsystematic research on shale gas in China and the blockage of core technologies and key tools by foreign companies, there was no method and key technology for quantitative shale gas assessment. As a result, the distribution pattern and the favorable zones and strata of shale gas resources were not clarified. In 2007, the Company conducted shale gas research together with the Newfield Exploration Company of the United States. In 2009, it implemented the first shale gas joint assessment project of China in the Fushun–Yongchuan area by cooperating with the Royal Dutch Shell Group. Through drilling and coring, profile observation, analysis and testing, and previous-data processing, the Company obtained the key parameter system for shale gas assessment of basins and established the methods for assessing geology and resources. Moreover, the Company established the technical method and quantitative index system for resource assessment and the selection of exploration strata and areas. Accordingly, the Wufeng–Longmaxi Formations were

determined as the most favorable exploration and exploitation strata at present, and the Changning, Weiyuan, Fushun–Yongchuan, and Zhaotong areas were selected as four favorable areas.

(2) **Pilot tests of shale gas exploration and exploitation (2009–2014):** To effectively produce shale gas in a basin with abundant resources, it is necessary to confirm the resources and assess the resource potential through the pilot tests of shale gas exploitation based on the first stage. Moreover, it is necessary to develop the main technologies for shale gas exploitation by tackling technological challenges. The purpose is to improve single-well production and establish efficient production management modes to improve efficiency and reduce costs. During the pilot test stage, a batch of horizontal wells was drilled and the main drilling and fracturing techniques were tested; a batch of platform wells was drilled, and the pilot tests of factory-pattern drilling and fracturing operations of horizontal wells were conducted; different roadway locations, spacings, orientations, and horizontal-section lengths of horizontal wells were designed; and exploitation technologies were optimized through tests. Through the pilot tests, the first shale gas well in China Wei 201 — was drilled and yielded gas through fracturing; the first horizontal well of shale gas in China Wei 201-H1 — was drilled, making a breakthrough in the drilling of horizontal wells and the large-volume fracturing technique; and the first shale gas horizontal well of commercial development value in China Ning-201-H1 — was drilled, making a breakthrough in the commercial development of shale gas. A series of technological breakthroughs strengthened the confidence of China in shale gas development and broke the foreign technological blockage, contributing to the preliminary determination of the main technologies for shale gas exploitation and the factory-pattern operations. Moreover, the lower part of the Long-$I_1$ Submember of the target of a horizontal well was designed as follows. The roadway spacing was 300–400 m and the trace direction was perpendicular to the maximum principal stress or intersected with the maximum principal stress and the fracture at large angles. Furthermore, efficient customized PDC bits and screws were used to accelerate drilling, oil-based drilling fluid was used to prevent collapse in shale sections, and

gamma-ray-guided drilling was adopted. The fracturing of horizontal wells was performed using cluster perforating and staged fracturing using cable-conveyed bridge plugs and the slug-type frac sand technique with low-viscosity slickwater and low-density medium-strength ceramsites. The factory-pattern drilling and fracturing operations characterized by "operating using double rigs, batch drilling, and standardized operations" and "integrated deployment, distributed fracturing, and alternating operations" were adopted to improve the operation efficiency and reduce costs. At this stage, a batch of appraisal wells was drilled and 2D and 3D seismic sounding was carried out to continue the resource and production capacity assessment in the Changning and Weiyuan areas.

(3) **Construction of national shale gas demonstration areas (2014–2016):** To accelerate the development of the shale gas industry, China approved the establishment of Changning and Weiyuan blocks as national shale gas demonstration areas in March 2012. The demonstration areas are designed to develop and improve the main technologies of shale gas exploitation, form the characteristic management modes, and build the annual production capacity of $20 \times 10^8$ $m^3$, aiming to achieve the effective exploitation of shale gas on a large scale and promote the development of the shale gas industry in China. Based on pilot tests, the China National Petroleum Corporation (CNPC) actively launched the construction of the two demonstration areas in 2014 as a response to the call of the state. It efficiently advanced the construction of demonstration areas by using its overall advantages and fully completed various tasks. Specifically, it built annual production capacity of $25 \times 10^8$ $m^3$, overfulfilling the task of production capacity construction; it achieved the proven shale gas reserves of $1{,}635.31 \times 10^8$ $m^3$ and confirmed the areas in southern Sichuan where shale gas can be exploited as well as the resource distribution in these areas; it mastered the technologies and means for effective shale gas exploitation, using which it improved the exploitation effects and increased the average output of shale gas wells during production tests from $11.1 \times 10^4$ $m^3/d$ to $21.9 \times 10^4$ $m^3/d$. Moreover, it established characteristic management systems and mechanisms and the factory-pattern operations, thus reducing

the single-well comprehensive cost from RMB 130 million yuan to RMB 50 million yuan; it fully promoted the health-safety-environment (HSE) system of production operations, achieving safe and clean production.

(4) **Industrial production (September 2016–present):** Through the construction of Changning and Weiyuan demonstration areas, the shale gas in southern Sichuan has been clearly assessed and the resources have been confirmed. Moreover, the technologies for shale gas production have become mature. Therefore, there are mature conditions for large-scale shale gas production. At present, the special 13th Five-Year Plan for the development of shale gas exploitation in the Sichuan Basin has been prepared. Moreover, the *Exploitation Plan of Annual Production of 50 × $10^8$ $m^3$ of the Changning Shale Gas Field* and the *Exploitation Plan of Annual Production of 50 × $10^8$ $m^3$ of the Weiyuan Shale Gas Field* have been prepared and were approved for implementation in October 2017. The current stage is witnessing a great effort made in promoting technological progress, management innovation, in-depth assessment, and large-scale production. The purpose is to achieve a greater development goal of shale gas exploitation. The Changning and Weiyuan blocks each had shale gas production of up to $50 \times 10^8$ $m^3$ in 2020. By the end of April 2018, a total of 73 horizontal shale gas wells were put into production in the Changning shale gas field, achieving daily gas production of $440.53 \times 10^4$ $m^3$, a maximum single-well production during production tests of $43.3 \times 10^4$ $m^3/d$, and the cumulative gas production over the years of $34 \times 10^8$ $m^3$.

### 2.2.2. *Difficulties in the seismic prediction of shale gas sweet spots*

Shale gas sweet spots refer to the shale gas reservoirs that have high TOC, large thicknesses, high gas content, microtectonic setting, developed fractures, rich brittle minerals, and high brittleness indices and can be easily fractured. They are the optimal areas or horizons for shale gas exploration and exploitation. Therefore, the prediction of sweet spots is the key to shale gas exploration and exploitation. However, shale gas exploration and exploitation in southern Sichuan have failed to yield satisfactory results due to the complexity of shale gas reservoirs themselves and the local shale gas reservoirs and tectonics. The prediction of sweet spots faces many technical problems in actual production, which mainly include the following:

(1) Despite thick and stably distributed shale strata and reservoirs, the sweet spots, which determine the exploitation effect of shale gas, are relatively thin vertically and strongly heterogeneous planarly. Therefore, it is difficult to accurately predict the sections with concentrated sweet spots and conduct fine-scale design of roadways.
(2) Ultramicropores with complex structure and bedding develop in shale gas reservoirs, for which there is no targeted seismic petrophysical modeling approach. Moreover, the geophysical response mechanisms and characteristics of shale gas reservoirs are not clear yet. They are to be determined through stematic and in-depth research to lay a foundation for the comprehensive prediction of sweet spots.
(3) The methods for predicting key parameters (e.g., TOC) have long relied on empirical formulas. Furthermore, the seismic inversion method itself has large limitations. Therefore, the prediction precision of the key parameters used to assess shale gas sweet spots cannot meet the demands for efficient shale gas exploitation.
(4) Limited by traditional seismic methods for predicting formation pressure, the seismic prediction of pore pressure presently suffers low precision and is inadequate to provide effective support for seismic prediction and engineering scheme design of overpressure sweet spots.
(5) Owing to various controlling factors and data sources of shale gas sweet spots, traditional comprehensive methods for oil and gas assessment only produce poor quantitative results when they are applied to shale gas assessment and cannot meet the current demands for shale gas exploitation.

### 2.2.3. *Technical measures for the seismic prediction of shale gas sweet spots*

Through the comprehensive study of the scientific and production problems to be solved and the in-depth analysis of the technical challenges in sweet spot prediction, the following technical measures have been developed.

(1) Conduct multi-factor comprehensive assessments by highly integrating the data on logging, geology, drilling, and production to determine the longitudinal distribution of sweet spots and provide a basis for roadway determination.
(2) Innovate the method for the seismic petrophysical modeling of shale gas reservoirs by fully considering the occurrence state of shale gas and reservoir pore structure, and establish a seismic petrophysical model. Conduct forward modeling using the seismic petrophysical model to determine the geophysical response characteristics of shale gas reservoirs.
(3) Select the optimal elastic parameters sensitive to TOC based on seismic petrophysical analysis, and establish the TOC quantitative prediction template by determining the quantitative relation between elastic parameters and TOC through seismic petrophysical simulation. Obtain high-precision elastic parameters by introducing the advanced pre-stack inversion of elastic parameters, and obtain high-precision TOC data under the constraints of the TOC quantitative prediction template.
(4) Improve and optimize traditional prediction methods and establish new seismic methods for pore pressure prediction by finding the root cause of the low accuracy of the existing seismic prediction method of pore pressure from the point of geomechanics. The purpose is to reduce the prediction errors of pore pressure and improve the prediction precision of formation pressure and overpressure sweet spots.
(5) Apply the advanced mathematical model to select the optimal factors that affect the distribution of shale gas sweet spots and sort them. Build a quantitative prediction model of shale gas sweet spots, quantitatively determine the weight of each key parameter for shale gas assessment in the final prediction results, and then associate the assessment results with the production results to achieve the quantitative prediction of sweet spots.

## Bibliography

Arcangelo S, Gabino C, Kevin C, *et al.* Seismic reservoir characterization in resource shale plays: "Sweet spot" discrimination and optimization of

horizontal well placement. *SEG Technical Program Expanded Abstracts*, 2011, 1744–1748.
Bian RK, *Natural Gas Accumulation Mechanism Spectra and Its Application*. Beijing: China University of Geosciences (Beijing), 2011.
Biot MA. General theory of 3D consolidation. *Journal of Applied Physics*, 1941, 12 (2): 155–164.
Castagna JP, Batzle ML, Eastwood RL. Relationships between compressional and shear-wave velocities in classic silicate rocks. *Geophysics*, 1985, 50 (4): 571–581.
Chalmers GR, Bustin RM, Power IM. Characterization of gas shale pore systems by porosimetry, pycnometry, surface area, and field emission scanning electron microscopy/transmission electron microscopy image analyses: Examples from the Barnett, Woodford, Haynesville, Marcellus, and Doig unit. *AAPG Bulletin*, 2012, 96(6): 1099–1119.
Cheng GS, Dong DZ, Wang SQ, *et al.* Analysis of controls on gas shale reservoirs. *Natural Gas Industry*, 2009, 29(5): 17–21.
Chorn L, Stegent N, Yarus J. Optimizing lateral lengths in horizontal wells for a heterogeneous shale play. *SPE/EAGE European Unconventional Resources Conference and Exhibition European Association of Geoscientists & Engineers*, 2014, 2014 (1): 1–12.
Chu HL, Tan CD, Song J. Comparison of the accumulation mechanisms and characteristics between natural gas, coalbed methane, and shale gas. *China Petroleum and Chemical Industry*, 2010, 9: 44–45.
Ding AX. *Characteristics and Log Evaluation Techniques of Shale Gas Reservoirs*. Qingdao: China University of Petroleum (East China), 2015.
Gassmann F. Elastic waves through a packing of spheres. *Geophysics*, 1951, 16 (4): 673–685.
Goodway B, Varsek J, Abaco C. Practical applications of P-wave AVO for unconventional gas Resource Plays—Part 2: Detection of fracture prone zones with Azimuthal AVO and coherence discontinuity. *CSEG Recorder*, 2006, 31(4). Available at: https://csegrecorder.com/articles/view/practical-applications-of-p-wave-avo-for-unconventional-gas-resource-plays.
Guo XS, Hu DF, Wen ZD, *et al.* Marine shale gas in the lower Paleozoic of Sichuan Basin and its periphery: A case study of the Wufeng-Longmaxi formations of Jiaoshiba area. *Geology in China*, 2014, 41(3): 893–901.
He XP, Gao YQ, Tang XC, *et al.* Analysis of major factors controlling the accumulation in normal pressure shale gas in the southeast of Chongqing. *Natural Gas Geoscience*, 2017, 28 (4): 654–664.
Jia CZ, Zheng M, Zhang YF. Four important theoretical issues of unconventional petroleum geology. *Acta Petrolei Sinica*, 2014, 35 (1): 1–10.
Jiang ZX, Li Z, Tang XL, *et al.* Main controlling factors of shale gas accumulation and preservation and the selection of optimal exploration targets. *Journal of Jilin University (Earth Science Edition)*, 2015 (1): 18–16.

Kong XX, Jiang ZX. Laminations characteristics and reservoir significance of fine-grained carbonate in the lower 3rd member of Shahejie Formation of Shulu sag. *Petroleum Geology and Recovery Efficiency*, 2016, 23(04): 19–26.

Langmuir I. The constitution and fundamental properties of solids and liquids. Part I. Solids. *Journal of the Franklin Institute*, 1917, 184 (5): 102–115.

Li DH, Li JZ, Wang SJ, *et al.* Analysis of the conditions of shale gas accumulation. *Natural Gas Industry*, 2009, 29 (5): 22–26.

Li YX, Nie HK, Long PY. Development characteristics of organic-rich shale and strategic selection of shale gas exploration area in China. *Natural Gas Industry*, 2009, 29 (12): 115–118.

Liu GD, Li J, Li JM, *et al.* The controls and the assessment method for the effectiveness of natural gas migration and accumulation process. *Natural Gas Geoscience*, 2005, 16 (1): 1–6.

Passey RQ, Greaney S, Kulla BJ, *et al.* Well log evaluation of organic rich rocks. In *14th International Meeting on Organic Geochemistry*, Paris, 1989, pp. 18–22.

Peake N, Castillo G, Van de Coevering N, *et al.* Integrating surface seismic, microseismic, rock properties and mineralogy in the Haynesville Shale. In *Unconventional Resources Technology Conference*, Denver, Colorado, 25–27 August 2014. Society of Exploration Geophysicists, American Association of Petroleum Geologists, Society of Petroleum Engineers, 2014, pp. 343–353.

Roy B, Hart B, Mironova A, *et al.* Integrated characterization of hydraulic fracture treatments in the Barnett Shale: The Stocker geophysical experiment. *Interpretation*, 2014, 2 (2): 111–127.

Snyder MJ, Schorr JR. *Characterization and Analysis of Devonian Shales as Related to Release of Gaseous Hydrocarbons.* Quarterly Technical Progress Report, March–May 1977, Battelle Columbus Labs, Columbus, OH. 1977.

Tinnin B, Bello H, Mcchesney M. Multi-source data integration: Eagle ford shale sweet spot mapping. *SPE/AAPG/SEG Unconventional Resources Technology Conference (URTEC)*, 2015. DOI:10.15530/urtec-2015-2154534.

Treadgold G, Mclain B, Sinclair S, *et al.* Eagle ford shale prospecting with 3D seismic data within a tectonicand depositional system framework. *Leading Edge*, 2011, 30 (1): 2270–2281.

Varga R, Lotti R, Pachos A, *et al.* Seismic inversion in the Barnett Shale successfully pinpoints sweet spots to optimize wellbore placement and reduce drilling risks. *Seg Technical Program Expanded Abstracts*, 2012: 4609. DOI:10.1190/segam 2012-1266.1.

Wang SQ, Wang SY, Man L, *et al.* Evaluation methods and key parameters for selecting shale gas exploration areas. *Journal of Chengdu University of Technology (Science and Technology Edition)*, 2013, 12 (6): 609–620.

Wu J, Jiang ZX, Wu MH, *et al.* Summary of research methods about the aequence stratigraphy of the fine-grained rocks. *Geological Science and Technology Information*, 2015, 34 (5): 16–20.

Xu, S, Saltzer RL, Keys, RG. Integrated anisotropic rock physics model. US, 2010. DOI: US20080086287 A1.

Yang ZH, Han ZY, Li MH, *et al.* Characteristics sand patters of shale gas accumulation in typical North American Cratonic basin sand their enlightenments. *Oil and Gas Geology*, 2013, 34 (4): 463–470.

Yuan XJ, Lin SH, Liu Q, *et al.* Lacustrine fine-grained sedimentary features and organic-rich shale distribution pattern: A case study of Chang 7 Member of Triassic Yanchang Formation in Ordos Basin, NW China. *Petroleum Exploration and Development*, 2015, 42 (1): 34–43.

Zeng QH, Qian L, Liu DH, *et al.* Organic petrological study on hydrocarbon generation and expulsion. *Acta Sedimentologica Sinica*, 2006, 24 (1): 113–122.

Zhang JC, Jin ZJ, Yuan MS. Reservoring mechanism of shale gas and its distribution. *Natural Gas Industry*, 2004, 24 (7): 15–18.

Zhao Q, Du D, Wang HY, *et al.* The characteristics analysis of the different types of shale gas reservoir. *Sino-Global Energy*, 2012, 17 (11): 43–47.

Zou CN, Yang Z, Zhu RK, *et al.* Progress in China's unconventional oil & gas exploration and development and theoretical technologies. *Acta Geologica Sinica*, 2015, 89 (6): 979–1007.

https://doi.org/10.1142/9789811283185_0003

# Chapter 3

# Characteristics and Seismic Responses of Shale Gas Sweet Spots

The Longmaxi Formation in southern Sichuan is the only stratum of China where large-scale shale gas exploitation has been achieved at present. Logging and seismic technologies are the most commonly used in the research on the longitudinal and transverse distribution of shale gas sweet spots. Therefore, analyzing the response characteristics of shale gas sweet spots reflected by log curves and seismic data is significant for shale gas prediction. This chapter first introduces the log response characteristics of shale gas sweet spots, determines the method for predicting the longitudinal distribution of sweet spots, and obtains the prediction results. Then, this chapter determines the elastic parameters sensitive to shale gas sweet spots through petrophysical analysis and modeling and, accordingly, establishes a seismic template for the quantitative prediction of shale gas reservoirs. Finally, this chapter determines the seismic response characteristics of shale gas sweet spots through the forward simulation using the seismic petrophysical model, thus laying a foundation for the prediction of shale gas sweet spots.

## 3.1. Log response characteristics of shale gas sweet spots

### 3.1.1. *Conventional log responses*

There are many similarities between shale gas and conventional gas. Both are non-conductive media with high natural gamma-ray (GR) values, high sonic interval transit time, low density, and low hydrogen index. The shale gas reservoirs in the Barnett Formation of the Newark East gas field show abnormally high natural gamma-ray log response, low density, and high resistivity, like other major shale gas producing layers in North America. These characteristics resemble the rock properties in conventional reservoirs in China. Organic-rich shales generally have fine-grinned sedimentary particles and high radioactive element content. Moreover, their rock density and sonic velocity decrease with increasing TOC, while their sonic interval transit time and resistivity increase with an increase in the organic matter content. Therefore, the changes in the physical properties and gas-bearing property parameters of shale gas reservoirs can be systematically studied using the comprehensive response characteristics of combined conventional log curves.

Figure 3.1 shows the lithology and the conventional log curves of the target horizons in the study area. Despite mainly consisting of shales, the Longmaxi Formation (including the Wufeng Formation) has complex lithology. Shales can be generally divided into black, calcareous, siliceous, carbonaceous, ferruginous, and oil shales. The shale gas reservoirs in the study area have complex lithology. Besides general shales, the target horizons show the development of mudstones, carbonaceous shales, silty mudstones, and siltstones. The complex lithology makes it difficult to identify shale gas reservoirs using conventional logging methods. Therefore, it is necessary to identify favorable shale gas reservoirs by analyzing the log response characteristics of different lithology and combining test and analytical results.

The Lower Silurian Longmaxi Formation is in conformable contact with the overlying Shiniulan Formation. This formation was deposited in a continental shelf environment, and its lithology mainly comprises mudstones and shales, with the content of clastics dominated by carbonaceous shale and siltstone shale gradually increasing toward its top boundary. The Shiniulan Formation was deposited in

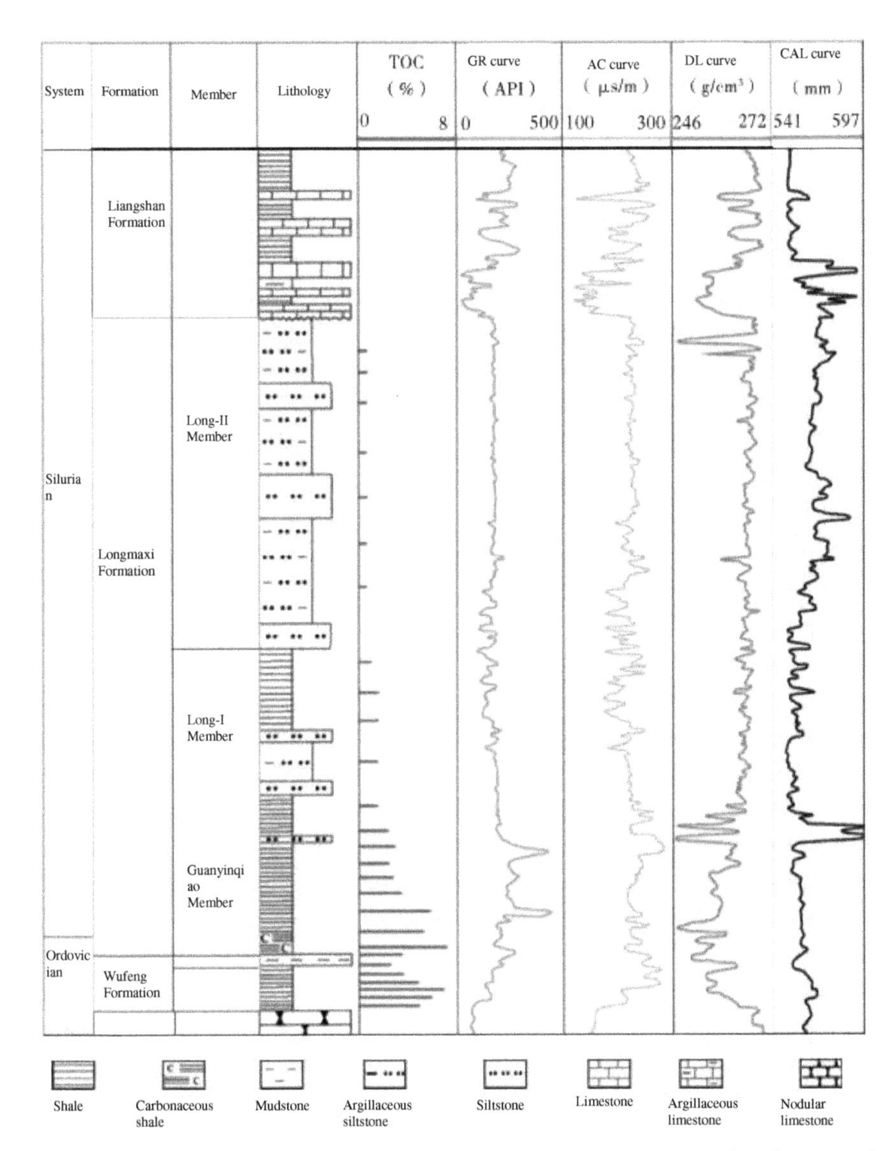


**Figure 3.1.** Conventional log response characteristics in the Longmaxi Formation.

a shallow-water carbonate platform. Its bottom boundary mainly comprises unequally thick interbeds consisting of calcareous siltstones and grayish-black mudstones, with calcareous components increasing in shallow water.

The natural gamma-ray (GR) curve of the Longmaxi Formation is flat at the top boundary and changes abruptly into a sawtooth shape upward from the Shiniulan Formation. For the sonic interval transit time or acoustic curve (AC), its values decrease approximately linearly from the Longmaxi Formation to the Shiniulan Formation except that they abruptly increase near the boundary between the two formations. The density log (DL) curve of the Longmaxi Formation is flat for the Longmaxi Formation, but its values gradually increase upward from the Shiniulan Formation. For the caliper log (CAL) curve of the Longmaxi Formation, its values gradually decrease at the top boundary of this formation and exhibit a bell-shaped distribution at the boundary between the Longmaxi and the Shiniulan formations. The TOC of the Longmaxi Formation shows a low-flat distribution with amplitude less than 1% near the boundary between the Longmaxi and Shiniulan formations.

The bottom boundary of the Longmaxi Formation is in comfortable contact with the Upper Ordovician Wufeng Formation throughout the Sichuan Basin, with the Guanyinqiao Member as their boundary. The Longmaxi Formation is a part of the Wufeng Formation in terms of stratigraphic division, and it mainly comprises shell limestones. The Wufeng Formation is composed of carbonaceous or siliceous shales in the Changning area and consists of limestones in the Weiyuan area. Therefore, this formation is a shale gas reservoir in the study area. However, it is thin in the study area, with the thickness greatly varying in the range of 2.3–5.3 m transversely. The Wufeng Formation in the entire Changning area thins from north to south. Specifically, it is approximately 10 m thick in the northern wing and becomes increasingly thin far away from the denudation area and is 0.5–15 m thick.

During the deposition of the Late Ordovician Wufeng Formation, the water level rose, and a set of deep-water siliceous shales was deposited in the Sichuan Basin. These shales, which contain large quantities of siliceous organisms and graptolites, are sediments of deep-water shelf facies. From the late sedimentary period of the Wufeng Formation to the early sedimentary period of the Longmaxi Formation, a brief interglacial period occurred globally, the sea level dropped rapidly, and a set of marls containing shallow-water Hirnantia-Dalmanitina biota (Guanyinqiao Member) were deposited in the basin. The interglacial period ended during the early deposition of the Longmaxi Formation, the sea level increased rapidly, and the basin was subjected to continuous napping from its periphery, leading to the wide deposition of a set of organic-rich shales in the Longmaxi Formation. The values of the GR curve change abruptly at the boundary between the Wufeng and Longmaxi Formations and peak at the bottom of the Longmaxi Formation. With the peak as a boundary, the strata with GR values greater than half the peak constitute the Longmaxi Formation, and those with GR values less than half the peak comprise the Wufeng Formation. The values of both AC and Density curves increase from bottom to top. The values of the TOC curve abruptly change in the form of a finger at the boundary and abruptly increase upward from the Longmaxi Formation.

During the deposition of the Longmaxi Formation, the sea level continuously dropped, which is reflected on the lithological profile. During this period, a short-duration transgression event occurred. Based on this event, the Longmaxi Formation can be divided into two secondary cycles, namely, the Long-I and Long-II members, the boundary between which is the gray silty shales at the bottom of the Long-I Member. The values of the GR curve continuously decrease from the top of the Long-I Member downward to the top of the boundary between the two members, showing a bell shape. The GR values increase rapidly briefly at the boundary between the two members due to the development of organic-rich shales and then continuously decrease downward from the boundary, exhibiting another bell shape. The values of the AC curve slightly increase from the Long-I Member to the Long-II Member, implying a slight decrease in sonic velocity. The sonic velocity in the Long-II member is slightly higher than that in the Long-I Member, which is closely related to the difference in the organic matter between the two members. The values of the DL curve oscillate in a spike pattern at the top of the Long-I Member and significantly increase from the Long-I Member to the Long-II Member. This variation trend is contrary to that of the TOC. The Long-I Member, which has TOC generally greater than 1%, serves as the main horizon where shale gas reservoirs are distributed in the study area.

The Long-I Member can be divided into two submembers, namely, Long-$I_1$ and Long-$I_2$ from bottom to top, according to the comprehensive response characteristics of its log curves. As shown in the

GR curve, the GR values of the Long-$I_1$ Submember are generally higher than those of the Long-$I_2$ Submember, and the GR values and their increased amplitude are increasingly high for the part closer to the bottom of the Longmaxi Formation. The AC curve of the Long-$I_2$ Submember shows a thick box shape, indicating stable sonic interval transit time and a sonic velocity. The DL curve shows a variation trend contrary to that of the AC curve. Specifically, the values of the DL curve increase in a funnel shape from the Long-$I_1$ to the Long-$I_2$ Submember, but rapidly decrease upward from the Long-$I_2$ Submember, showing a box shape. Overall, the values of the DL curve of the Long-$I_2$ Submember are higher than those of the Long-$I_1$ Submember. TOC can also significantly distinguish the Long-$I_1$ Submember from the Long-$I_2$ Submember. TOC becomes 2% (indicating high-quality shale interval) downward from the boundary. The TOC of the Long-$I_1$ Submember decreases upward in a bell shape, while the TOC of the Long-$I_2$ Submember is less than 2% with little variation.

The Long-$I_1$ Submember and the Wufeng Formation have high TOC and are the main target horizons of shale gas in southern Sichuan. They are, therefore, the focus and important topics of present shale gas exploration and exploitation, as well as the focus of current geological and engineering technologies. Studies have shown that it is not scientific to simply divide the Long-$I_1$ Submember into two upper and lower Long-$I_1$ sublayers since this division cannot meet the requirements for large-scale shale gas exploitation. Therefore, according to the production requirements, the Long-$I_1$ Submember is subdivided into four sublayers, namely, Long-$I_1^1$, Long-$I_1^2$, Long-$I_1^3$, and Long-$I_1^4$ from bottom to top, as shown in Figure 3.2.

(1) **Long-$I_1^4$ Sublayer:** A clear lithologic boundary exists between the Long-$I_1^4$ and Long-$I_1^3$ sublayers. It consists of a set of black-gray silty shales at the bottom of Long-$I_1^4$ and a set of grayish-black calcareous shales at the top of Long-$I_1^3$. The Long-$I_1^4$ Sublayer was deposited as a reversed cycle formed by a slow regression. The GR curve of the Long-$I_1^4$ Sublayer is relatively low and stable, with average GR values of 120–150 on the American Petroleum Institute (API) gravity scale. The resistivity logging (RL) curve tends to decrease slightly upward from the boundary between Long-$I_1^4$ and Long-$I_1^3$, with an average value of approximately 25 $\Omega\cdot$m. The values of the AC curve decrease from

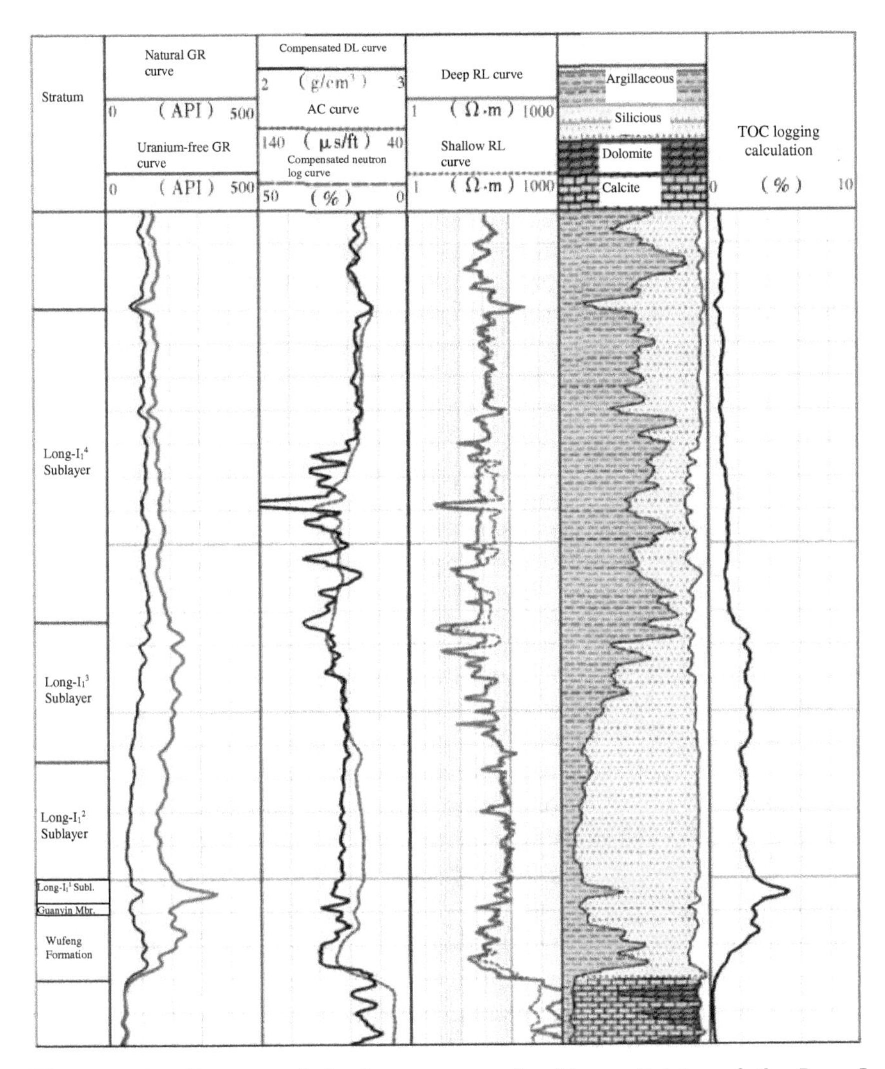


**Figure 3.2.** Features of the log curves and sublayer division of the Long-I Submember.

*Note*: Mbr. refers to Member, and Subl. refers to Sublayer.

Long-$I_1^3$ to Long-$I_1^4$, while the values of the DL crave decrease at the boundary and then gradually increase from Long-$I_1^3$ to Long-$I_1^4$, with an average of approximately 2.56 g/cm$^3$. Regarding mineral components, the Long-$I_1^4$ Sublayer has the lowest carbonate content of only 25% among the four sublayers but has high clay mineral and siliceous contents of up to approximately 40%.

(2) **Long-$I_1^3$ Sublayer:** The boundary between the Long-$I_1^3$ and Long-$I_1^2$ Sublayers is the boundary between the black carbonaceous shale at the top of Long-$I_1^2$ and the grayish-black silty calcareous argillaceous shale at the bottom of Long-$I_1^3$. The Long-$I_1^3$ Sublayer comprises sediments of calcareous muddy shelf facies formed by slow regression and mainly consists of grayish-black calcareous shales. Among the four sublayers, Long-$I_1^3$ has the most developed graptolites, which show the highest quantity and the largest number of types, and is represented by the curved-back crown graptolites. As shown by the particle size of thin sections observed using a microscope, the biggest difference between Long-$I_1^3$ and Long-$I_1^4$ is that Long-$I_1^3$ bears fewer clay–silty sands and more clay particles, while both sublayers have roughly the same silty sands. The Long-$I_1^3$ Sublayer in the Changning area thickens from west to east, with a thickness of 6–16 m. The values of the GR curve abruptly change at the boundary between Long-$I_1^3$ and Long-$I_1^2$. Moreover, they are roughly stably distributed within Long-$I_1^3$ and show a box shape, ranging from 140 API to 180 API (average: 160 API). The values of the RL curve increase slightly from Long-$I_1^2$ to Long-$I_1^3$ and slightly fluctuate at the top of Long-$I_1^2$. The values of the AC curve decrease significantly upward from the boundary between Long-$I_1^2$ and Long-$I_1^3$. The values of the DL curve gradually increase upward from Long-$I_1^3$, with an average density of approximately 2.56 g/cm$^3$. The carbonate rocks have a mineral content of approximately 10%–20% and low clay content generally less than 30%.

(3) **Long-$I_1^2$ Sublayer:** The Long-$I_1^2$ Sublayer has a stable thickness and lithological distribution. It can be compared throughout the whole study area and is a marker layer in the whole area. Long-$I_1^2$ is mainly composed of black carbonaceous shales, which are sediments of carbonaceous muddy shelf facies formed by continuous slow progradation. Regarding sedimentary cycles, Long-$I_1^2$ is a retrogradation-type positive cycle. This sublayer has many types of graptolites dominated by diplograptus and glyptograptus, contains calcareous nodules and pyrite, and shows the development of shale bedding. It has a relatively stable thickness throughout the study area, which is usually approximately 6 m and does not exceed 10 m. The values of the GR curve corresponding to Long-$I_1^2$ and Long-$I_1^1$ exhibit different response

characteristics. Specifically, they increase in a bell shape from Long-$I_1^1$ to Long-$I_1^2$. They fluctuate greatly in a gyroscopic form within Long-$I_1^2$, ranging from 160 to 270 API (average: 200 API). The resistivity of Long-$I_1^2$ slightly decreases. The AC curve shows a similar variation trend to that of the GR curve. Long-$I_1^2$ has low content of brittle minerals (e.g., siliceous minerals) of generally 40%, increased carbonate rock content of approximately 15%, and roughly consistent clay mineral content.

(4) **Long-$I_1^1$ Sublayer:** Long-$I_1^1$ is located at the bottom of the Longmaxi Formation. During the disposition of the Long-$I_1^1$ Sublayer, the sea level was the highest compared with the other three sublayers and was a progradation-type reversed cycle formed under a slow regression. It mainly consists of black carbonaceous shales and has carbonaceous muddy shelf facies overall. Long-$I_1^1$ contains climacograptus, glyptograptus, and calcareous nodules. Its lithology is significantly different from that of the Guanyinqiao Member at the bottom, and there is a distinct biological boundary between them. However, there is no noticeable lithological boundary between the top of Long-$I_1^1$ and the bottom of Long-$I_1^2$. Long-$I_1^1$ has a stable thickness of 10–13 m and unusually high GR values of 200–500 API, which gradually decrease upward, form a bell-shape, and show the maximum value at the bottom. Long-$I_1^1$ has stable and high resistivity, showing a small increased trend upward and an average of approximately 100 $\Omega \cdot$ m. The DL curve of this sublayer has significant characteristics and can also be used to identify Long-$I_1^1$ within the study area. Specifically, Long-$I_1^1$ has the lowest density among the four sublayers in the Long-$I_1$ Submember; and the density of Long-$I_1^1$ shows an inverse finger shape and varies in the range of 2.1–2.5 g/cm$^3$. Long-$I_1^1$ has carbonate rock content of 0%–30%, brittle mineral content of approximately 50%, clay content of approximately 20%, and high pyrite content roughly greater than 4%.

The four sublayers have a consistent varying trend regarding thicknesses and thicken toward the north. Long-$I_1^1$ has a thickness of 1.3–2.4 m, with an average of 1.9 m. Long-$I_1^2$ has a thickness of 4.6–11 m, with an average of 8.8 m. Long-$I_1^3$ has a thickness of 3.5–7.5 m, with an average of 5.0 m. Long-$I_1^4$ has a thickness of 6.4–27 m, with an average of 15.7 m.

### 3.1.2. *Image log response*

Another important characteristic of shales is the development of natural fractures and bedding. Fractures in shale mainly take the form of shale bedding fractures and open or closed tectonic fractures. The assessment of shale gas fractures mainly has two purposes. The first purpose is to understand the development degree of shale bedding fractures in organic-rich shale intervals since the development of shale bedding fractures can effectively increase the storage space of free gas. The second purpose is to assess the preservation conditions of shale gas. Image log can be used to assess the development degree and the quantity of tectonic fractures in shale gas horizons and determine the presence of calcite veins in these tectonic fractures. Considering the low resistance background of shale strata, the data on resistivity image logs and resistivity can be used to effectively identify shale bedding fractures and high-resistance fractures. Figure 3.3 shows the high-resistance fractures, faults, and high-conductivity fractures identified in the Longmaxi Formation based on image log data. The high-resistance fractures are present as a bright sine curve, faults are shown as a dark sine curve with a wide dark area, and the high-conductivity fractures are shown as a dark sine curve in the form of a narrow strip.

The formation microimaging (FMI) image log data show the uneven longitudinal distribution of fractures. High-angle fractures are widely developed in the Wufeng Formation but are not developed in other strata, where only low-angle fractures and beddings have developed. Horizontal fractures are widely developed longitudinally. However, they are concentrated in the Wufeng Formation, especially in the Long-$I_1$ Submember.

As shown by the processing results of resistivity image logs and core data of the shale reservoirs at the location of well Ning-203, the fractures mainly consist of tectonic fractures and overpressure fractures, which are mostly filled with calcite and quartz minerals. FMI image log of this well shows that the shale horizon of the Wufeng Formation and the Long-I Member of the Changning shale gas field generally have no fracture overall. Some wells reveal a small number of fractures in the Long-$I_2$ Submember. For instance, well Ning-203 reveals that an open fracture has developed in Long-$I_2$ Submember (Figures 3.4 and 3.5).

As shown in Figure 3.5, the fractures in the Changning shale gas field are dominated by high-conductivity fractures, followed by a few

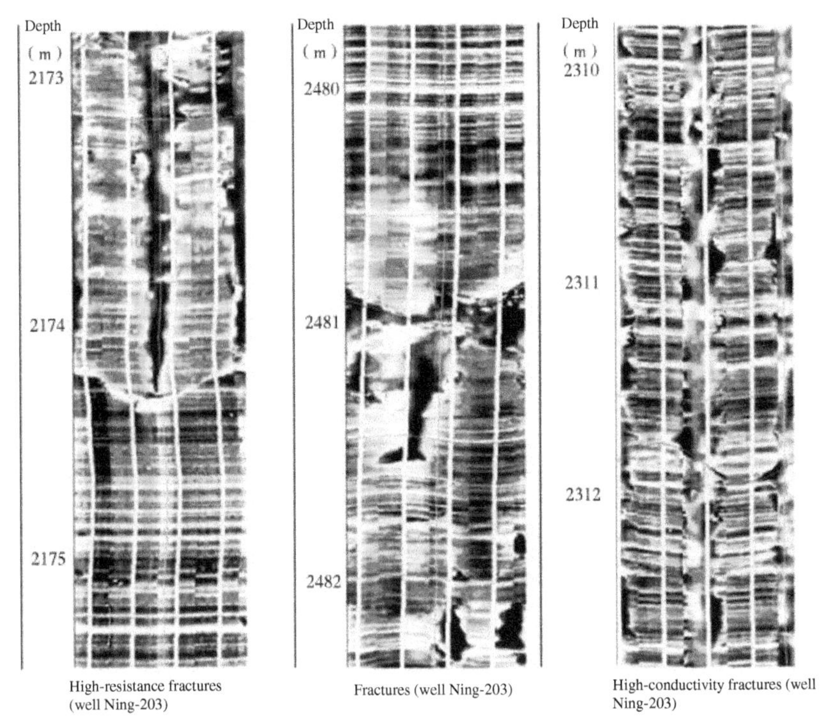


**Figure 3.3.** Fracture characteristics in image log of the shale interval of the Longmaxi Formation.

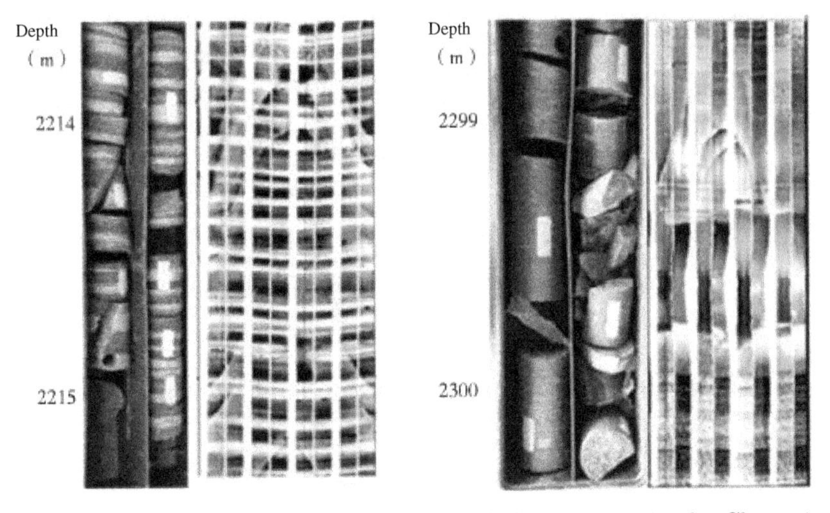


**Figure 3.4.** Characteristics of fractures in shale reservoirs in the Changning shale gas field.

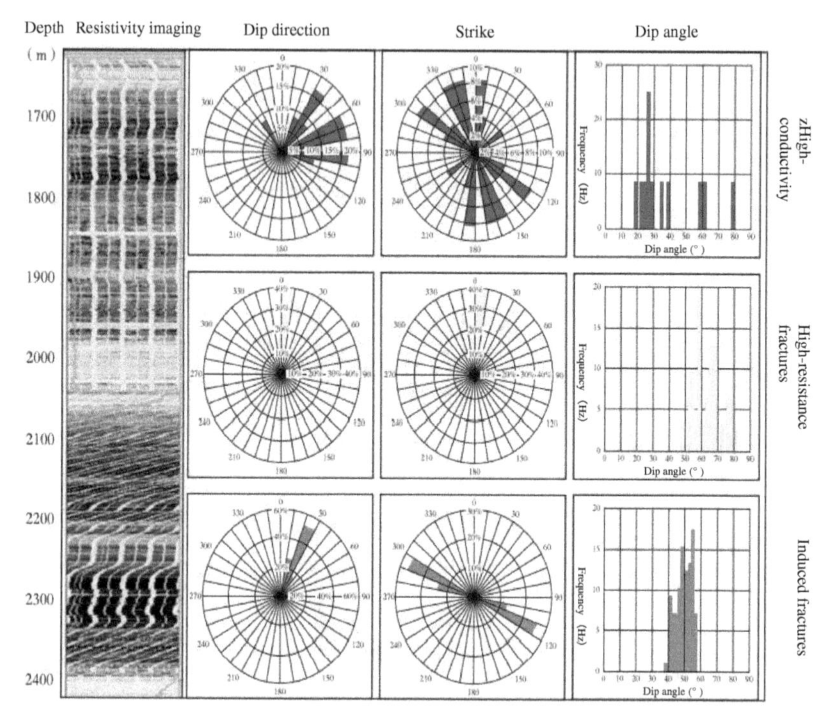


**Figure 3.5.** FMI image log-revealed fracture characteristics of the Wufeng Formation — The Long-I Member at the location of well Ning-203.

high-resistance and induced fractures. The high-conductivity fractures have a dip direction of 0–90° and an NW–SE strike in general. These numerous high-conductivity fractures provide space for natural gas storage and facilitate the formation of a complex fracture network in the late fracturing stimulation, thus achieving volume stimulation.

## 3.2. Methodology for predicting the longitudinal distribution of shale gas sweet spots

### 3.2.1. *Calculation of important geological parameters of reservoirs*

(1) **TOC:** The methods used to calculate TOC from well log data mainly include the $\Delta\log R$ method, the elemental capture spectroscopy (ECS) method, multimineral modeling, and the

regression method based on the relationship between radioactive uranium content, density, and TOC. Based on the measurement of the radioactive element contents in shales, the experimental analysis of TOC in shales, and the comparison of existing models, this book employs the $\Delta\log R$ method to estimate TOC from well log date due to its high applicability in practical assessment.

The $\Delta\log R$ method was proposed by Passey in 1989 and has been developed through research and modification based on massive test results by Exxon/ESSO since 1979. This method has been successfully applied to many wells around the world. It is highly applicable to both carbonates and clastic rocks, and can be used to accurately predict TOC.

$$\Delta\log R = \lg(R/R_{\text{baseline}}) + K(\Delta t - \Delta t_{\text{baseline}}) \tag{3.1}$$

$$\text{TOC} = (\Delta\log R) \cdot 10^{2.297-0.1688\text{LOM}} \tag{3.2}$$

where $\Delta\log R$ is the amplitude difference between the porosity curve (e.g., the AC curve) and the RL curve that has been scaled using core data; $R$ is the resistivity of shale gas reservoirs; $R_{\text{baseline}}$ is the resistivity of non-source rocks; $\Delta t$ is the sonic measurement; $\Delta t_{\text{baseline}}$ is the sonic corresponding to the $R_{\text{baseline}}$ of lean shale; and $K$ is the scaling factor, which depends on the unit of porosity logging. Moreover, TOC is the total organic carbon content, and LOM is the level of thermal maturity, which is a function of vitrinite reflectance ($R_o$).

When the $\Delta\log R$ method was employed to calculate the TOC of the Wufeng–Longmaxi Formations at the location of well Ning-201, the parameters were set to $R_{\text{baseline}} = 6.5$, $\Delta t_{\text{baseline}} = 85$, and LOM $= 12$. Figure 3.6 shows a comparison between the results calculated using the $\Delta\log R$ method and the core test results. The blue curve in the last column represents the TOC calculated based on well log data, and short red lines are the TOC measured from core data. The comparison shows that these two sets of results agree well, indicating that the $\Delta\log R$ method can be used to accurately calculate the TOC of the shale reservoir in the study area.

(2) **Porosity:** Unlike conventional sandstone and limestone reservoirs, shale gas reservoirs have poor physical properties and extremely low porosity and permeability. In addition, shale gas reservoirs have more complex pore types, which mainly include

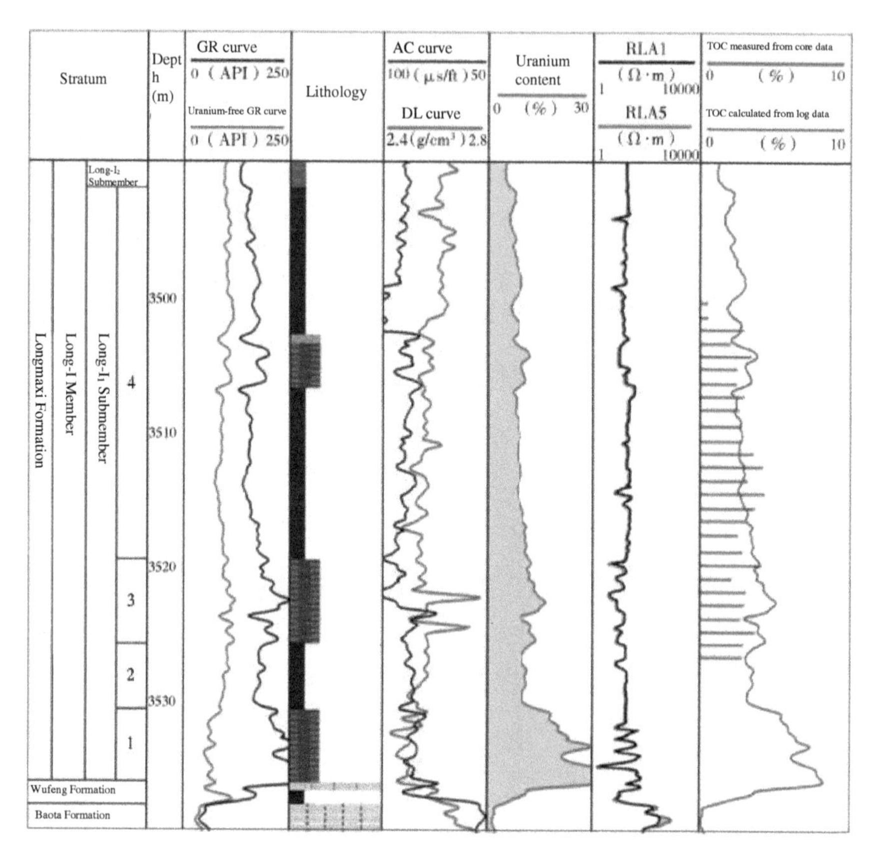


**Figure 3.6.** Comparison between the TOC measured from core data and that calculated based on well log data of the Wufeng–Longmaxi Formations.

nanopores in organic matter, the intercrystalline and interlayer pores of clay minerals, and intergranular pores of minerals such as quartz or feldspar, as well as the pore spaces of microfractures such as shale bedding fractures.

Since shale gas reservoirs have low porosity and thus generally contain no free movable water, the calculated water saturation is considered the irreducible water saturation. The multi-mineral model developed by Schlumberger can be used based on relevant data. Then, the non-correlation of the equation matrix in the model is minimized by adjusting its input parameters and employing optimization techniques. As a result, the volumes of various minerals and fluids are determined. The optimized method enjoys the biggest advantage that it can solve multiple models simultaneously. Specifically, the system can optimize the probabilistic combination of multiple models, including the lithology, fluid, and porosity models. Then, it can determine the model combination with the highest probability and reservoir parameters. The log response equations are as follows:

$$\emptyset_n = \emptyset \cdot [A(1 - S_{\text{xo}})\Delta T_{\text{hr}} + S_{\text{xo}}\Delta T_{\text{mf}}] + V_{\text{sh}}\Delta T_{\text{sh}} + \sum_{i-1}^{n} V_{\text{mai}}\Delta T_{\text{mai}} \quad (3.3)$$

$$\emptyset_n = \emptyset \cdot [A(1 - S_{\text{xo}})\emptyset_{\text{Nhr}} + S_{\text{xo}}\emptyset_{\text{Nmf}}] + V_{\text{sh}}\emptyset_{\text{Nsh}} + \sum_{i=1}^{n} V_{\text{mai}}\emptyset_{\text{Nmai}} \quad (3.4)$$

$$P_b = \emptyset \cdot [A(1 - S_{\text{xo}})P_{\text{hr}} + S_{\text{xo}}\rho_{\text{mf}}] + V_{\text{sh}}\rho_{\text{sh}} + \sum_{i=1}^{n} V_{\text{mai}}\rho_{\text{mai}} \quad (3.5)$$

$$\phi + V_{\text{sh}} + \sum_{i=1}^{n} V_{\text{mai}} = 1 \quad (3.6)$$

where $\Delta T_{\text{hr}}$, $\Delta T_{\text{mf}}$, $\Delta T_{\text{sh}}$, $\Delta T_{\text{ma}}$, and $\Delta T$ denote the sonic interval transit time of residual natural gas, mixed fluids, clay, and rock matrix, and the sonic interval transit time curve, respectively, $\mu$s/ft; $\varphi_{\text{Nhr}}$, $\varphi_{\text{Nmf}}$, $\varphi_{\text{Nsh}}$, $\varphi_{\text{Nma}}$, and $\varphi_n$ denote the neutron values of residual natural gas, mixed fluids, clay, and rock matrix, and the neutron value curve, respectively, %; and $\rho_{\text{hr}}$, $\rho_{\text{mf}}$, $\rho_{\text{sh}}$, $\rho_{\text{ma}}$, and $\rho_b$ are the density of residual natural gas, mixed fluids, clay, and rock matrix, and the density curve, respectively, g/cm$^3$. Moreover, $\Phi$ is the effective porosity, %; $S_{\text{xo}}$ is the residual gas saturation in the flushing zone, %; $V_{\text{mai}}$ is the mineral volume content, %; and $A$ is a dimensionless constant and is mainly determined through petrophysical experiments.

The square error equation for a single DL curve is as follows:

$$\varepsilon_{\rho b}^2 = \left(\frac{\rho_{b-}f_{\rho b}}{U_{\rho b}}\right)^2 \quad (i = 1, n_{\text{vol}}) \quad (3.7)$$

where $\varepsilon_{\rho b}$ is the error between the measured and predicted values of density; $\rho_b$ is the measured value of density; $f_{\rho b}(\upsilon_i)$ is the predicted value of density, which is a function of the volume of the target strata; and $U_{\rho b}$ is the uncertainty of density.

The above equation also applies to other parameters. Adding the square errors for each response equation yields the expression of the least square-error sum of all equations, as follows:

$$\Delta^2 = \sum_{k=1}^{n_{\text{tool}}} \varepsilon_k^2 \tag{3.8}$$

Equation (3.8) is the objective function used to determine the optimal solution of the response equations and is also called the non-correlation function. It is used to calculate the correlation or consistency between the measured values and the predicted values of the petrophysical model. When the non-correlation of the equation matrix is minimized, the volume and porosity of the rock minerals in the strata are obtained.

Figure 3.7 shows the comparison of porosity calculated from core analysis and log calculation of well Ning-201. The blue curve in the last column represents the porosity calculated based on well log data, and short red lines denote the porosity measured from core data. The comparison shows that these two sets of results agree well, indicating that the optimized method can be used to accurately calculate the porosity of the shale reservoir in the study area.

(3) **Total gas content:** Shale gas has complex occurrence states. It is generally believed that shale gas consists primarily of free gas and adsorbed gas, and some experts consider that it also includes water-dissolved gas. The adsorbed gas is closely related to TOC, while the free gas is closely related to porosity.

Studies show that there are many factors controlling the adsorption capacity of shale gas reservoirs, including clay mineral components and their contents, organic matter maturity, the geothermal gradient of shale gas reservoirs, reservoir pore pressure, water saturation, and shale gas components. The amount of adsorbed gas is proportional to the organic carbon content, the formation pressure, and the geothermal gradient. In general, higher pressure corresponds to higher gas content, and higher temperature and geothermal gradient are associated with more free gas and less adsorbed gas in the reservoirs.

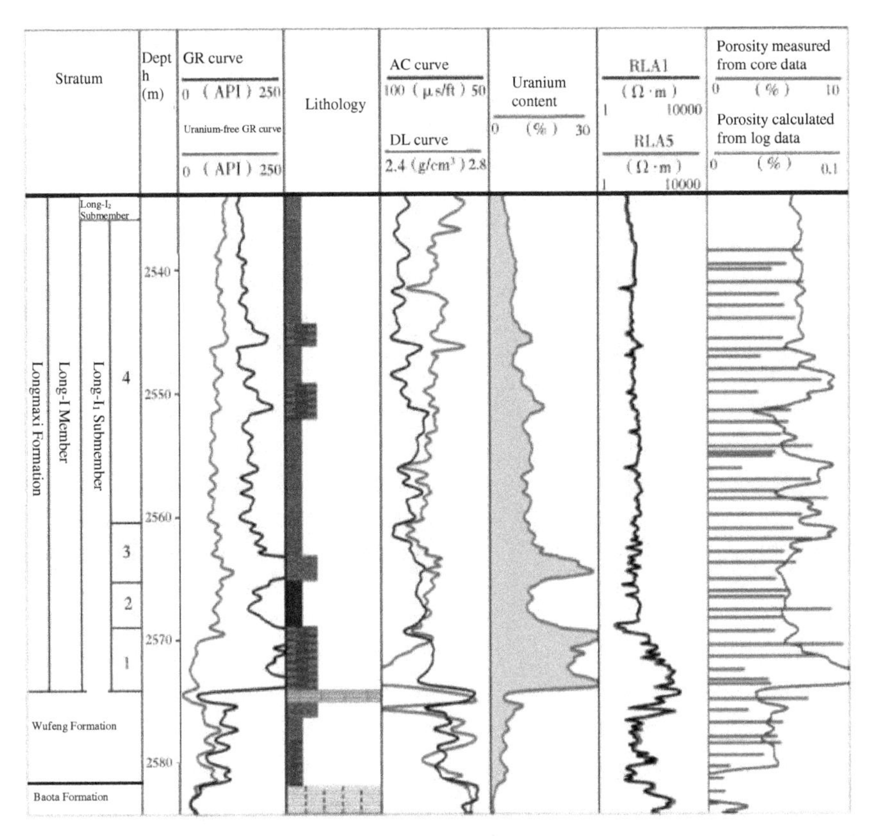


**Figure 3.7.** Comparison between the porosity measured from core data and that calculated based on well log data of the Wufeng–Longmaxi Formations.

The gas adsorption of shale satisfies the Langmuir equation, in which the parameters are determined according to the following isothermal adsorption experiments:

$$V = \frac{V_L p}{p + p_L} = \frac{V_L kp}{1 + kp} \tag{3.9}$$

where $V$ is the volume of shale in the standard state; $V_L$ is the volume of shale in the adsorption state; $p$ is the standard atmospheric pressure; $p_L$ is the adsorption pressure; and $x = \frac{1}{pL}$ is the adsorption coefficient, which is related to the adsorbent properties and represents the capacity of the solid for adsorbing gas.

The logging evaluation of the adsorbed gas content in shale is mainly achieved by determining the Langmuir equation of the shale region through isothermal adsorption experiments and then calculating the adsorbed gas content in strata by combining the temperature and the pressure of the strata.

Compared with the adsorbed gas content, the free gas content in shale gas reservoirs can be calculated simply. Since it is mainly controlled by the water saturation and the effective porosity of reservoirs, the calculation method for it is similar to that of gas saturation in conventional natural gas reservoirs, except that the gas content rather than the gas saturation should be calculated for the final evaluation of shale gas reservoirs. Moreover, the calculated gas content under reservoir conditions should be converted into that under the standard conditions of 25°C and 1 atmosphere. The final free gas content should be the volume of free gas per ton of rocks under standard conditions. Therefore, the calculated free gas content will be affected by reservoir pore pressure, formation temperature, and the compressibility factor of natural gas.

The free gas content under reservoir conditions can be converted into that under standard conditions (25°C and 1 atmosphere) using the following equation derived from the material balance equation of gas:

$$V_f = \frac{Q_f p_{\log}(25 + 273)}{P_0(T_{\log} + 273)Z} \tag{3.10}$$

where $V_f$ is the free gas content, m3/t; $p_{\log}$ and $T_{\log}$ are the formation pressure, Mpa, and temperature, °C, respectively; $P_0$ is 1 atmosphere, 0.1013 MPa; and $Z$ is the original gas deviation factor of a gas reservoir and can be obtained from the actual high-pressure physical properties or the gas component analysis of various wells.

The total gas content of the reservoir at a certain depth is calculated as follows:

$$V_t = V_a + V_f \tag{3.11}$$

where $V_t$ is the total gas content, $m^3/t$; $V_a$ is the corrected adsorbed gas content, $m^3/t$; and $V_f$ is the free gas content, $m^3/t$.

The above methods were applied to calculate the adsorbed gas content and the free gas content of shale gas reservoirs under different

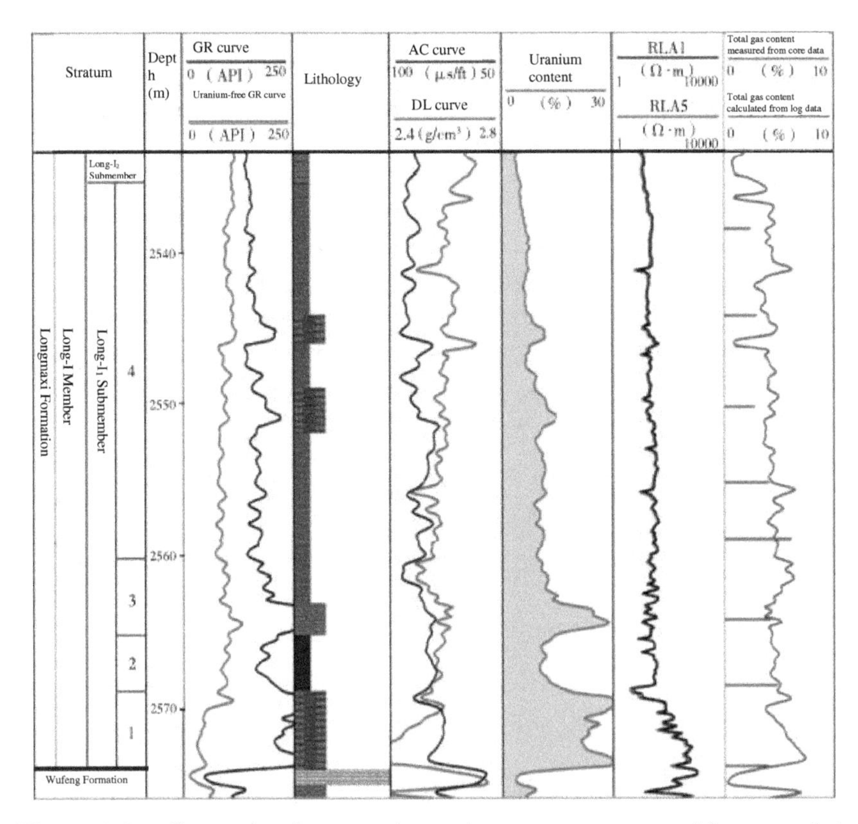


**Figure 3.8.** Comparison between the total gas content measured from core data and that calculated based on well log data of the Wufeng–Longmaxi Formations.

formation pressures and pore pressures. Then, the total gas content was calculated, as shown in Figure 3.8, in which the blue curve in the last column represents the total gas content calculated based on well log data, and short red lines denote the total gas content measured from core data. The comparison shows that these two sets of results agree well. Therefore, the method proposed in this book can meet the requirements of the calculation of the total gas content.

### 3.2.2. *Calculation of key engineering parameters of reservoirs*

The brittleness evaluation distinguishes the evaluation of shale gas reservoirs from that of conventional gas reservoirs. Shale brittleness is an important parameter for evaluating the feasibility of volume fracturing of shale gas reservoirs, the design and optimization of a volume fracturing scheme, the selection of the fluid system, post-fracturing fracture morphology, and the evaluation of post-fracturing capacity.

The brittleness of shale gas reservoirs is mainly affected by factors such as mineral composition and the development degree of shale bedding. It is a constant and is generally characterized by the brittleness index. However, shale features anisotropy and has shale bedding, which greatly changes the rock mechanical properties of shales. Therefore, it is not sufficient to evaluate the feasibility and effect of fracturing only using the brittleness index.

The current log evaluation of the brittleness of shale gas reservoirs mainly uses rock mechanical models, mineral content models, and composite models consisting of elastic parameters and mineral contents. Among them, the mineral content models are widely used in the log evaluation of shale brittleness as the well logging can more accurately recognize mineral contents.

Current techniques for evaluating the brittle mineral contents of shale gas reservoirs rely on the ECS technique and the photoelectric factor (PeF) curve of bulk density developed by Schlumberger. In this book, both methods are tried and the evaluation results using these two methods are compared with the results of X-ray diffraction analysis. As shown in Figure 3.9, the point-by-point comparison demonstrates that evaluation results using these two methods agree well with the X-ray diffraction analysis results. According to these results, the target horizons have high brittle mineral content (above 50%) and high fracability overall.

### 3.2.3. *Longitudinal distribution of shale gas sweet spots in southern Sichuan*

As stipulated in Z/T 0254-2020 *Shale Gas Resource and Reserve Estimation Specifications* issued by the Ministry of Natural Resources of the People's Republic of China, the parameters for shale gas reservoir evaluation include the reservoirs' effective thickness, total gas content, TOC, $R_o$, and brittle mineral content (Table 3.1); the lower limits of gas-bearing shale are defined as TOC $\geq$1%,

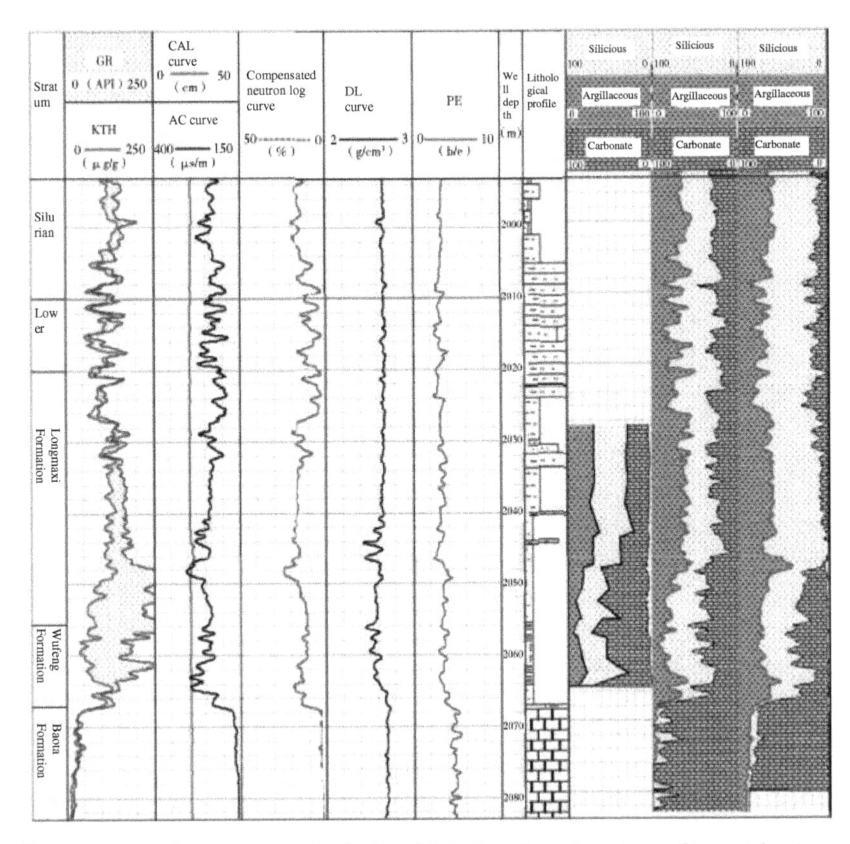


**Figure 3.9.** Comparison of calculated brittle mineral content of target horizons.

**Table 3.1.** Evaluation parameters of shale gas reservoirs.

| Effective shale thickness (m) | Total gas content ($m^3/t$) | TOC (%) | $R_o$ (%) | Brittle mineral content (%) |
|---|---|---|---|---|
| ≥50 | 1 | | | |
| 30–50 | 2 | 1 | 0.7 | 30 |
| <30 | 4 | | | |

$R_o \geq 0.7\%$, and brittleness ≥30%; and the lower limit of the total gas content varies with effective shale thickness and is ≥1 $m^3/t$, ≥2 $m^3/t$, and ≥4 $m^3/t$ for effective shale thickness of ≥50 m, 30–50 m (excluding 50 m), and <30 m, respectively.

**Table 3.2.** Classification criteria for shale reservoirs.

| Parameter | Shale reservoirs | | |
|---|---|---|---|
| | **Class I** | **Class II** | **Class III** |
| TOC (%) | ≥3 | 2–3 | 1–2 |
| Effective porosity (%) | ≥5 | 3–5 | 2–3 |
| Brittleness (%) | ≥55 | 45–55 | 30–45 |
| Total gas content ($m^3/t$) | ≥3 | 2–3 | 1–2 |

Since the effective porosity of shale greatly affects the free gas content in the total gas, porosity is used as one of the shale reservoir evaluation indicators. This study determined the classification criteria for the marine shale reservoirs in the Wufeng–Longmaxi Formations in the Sichuan Basin by integrating the classification criteria for reservoirs in various major shale gas fields at home and abroad. Specifically, the shale reservoirs in the Wufeng–Longmaxi Formations were divided into Class-I, -II, and -III reservoirs (Table 3.2) according to four selected geological parameters, i.e., TOC, gas content, effective porosity, and brittle mineral content. According to the determined criteria, Class-I reservoirs must have TOC ≥3%, total gas content ≥3 $m^3/t$, effective porosity ≥5%, and brittle mineral content ≥55%. Class-I and -II reservoirs must have TOC ≥2%, total gas content ≥2 $m^3/t$, effective porosity ≥3%, and brittle mineral content ≥45%.

According to the framework of the above classification criteria, the characteristics and distribution pattern of each member, submember, and sublayer of the reservoir in the study area are summarized by integrating lithology, sedimentology, paleontology, and electrical properties. Based on the log response, the physical and electrical properties, and the brittleness of reservoirs, the Long-I Member can be divided into two submembers, namely, Long-$I_1$ and Long-$I_2$. Moreover, the Long-I Member can be further divided into four sublayers, namely, Long-$I_1^1$, Long-$I_1^2$, Long-$I_1^3$, and Long-$I_1^4$, which have significantly different characteristics (Table 3.3).

The lithology of the Long-$I_1$ and Long-$I_2$ Submembers is divided by the dark gray shales at the bottom of the Long-$I_2$ Submember and their underlying grayish-black shales in the Long-$I_1$ Submember. Compared to the Long-$I_2$ Submember, the Long-$I_1$ Submembers

**Table 3.3.** Characteristics of the sublayers of the Long-$I_1$ Submember of the Longmaxi Formation in the Weiyuan area.

| | Petrological characteristics | | | | Paleontological characteristics | | Electrical characteristics | | | |
|---|---|---|---|---|---|---|---|---|---|---|
| Sublayers | Lithology | Colors | Special minerals | Sedimentary tectonic characteristics | Graptolite genus | Graptolite quantity | Natural gamma ray (API) | Sonic interval transit time | Resistivity | Density |
| Long-$I_1^1$ | Carbonaceous shales, silicious shales | Black | Pyrite | Horizontal bedding, calcareous strips, disturbance | Orthograptus, climacograptus, amplexograptus, petalolithus | Large quantity, more species | 121.00–671.00 | 66.50–149.00 | 1.10–638.00 | 2.15–2.79 |
| Long-$I_1^2$ | Carbonaceous shales, mudstones | Black, grayish-black | Pyrite | Horizontal bedding, vermicular distribution of pyrite, asphalt scratches | Glyptograptus, orthograptus, climacograptus, pristiograptus, akidogroptus, didymograptus | Large quantity, many species | 90.00–242.00 | 71.00–140.00 | 2.00–63.00 | 2.20–2.69 |
| Long-$I_1^3$ | Argillaceous shales, carbonaceous mudstone | Grayish-black, black | Pyrite | Horizontal bedding, cross bedding, nodules | Glyptograptus, climacograptus, diplograptus, didymograotus | Large quantity, more species | 102.00–362.00 | 73.00–149.00 | 4.00–50.00 | 2.36–2.87 |
| Long-$I_1^4$ | Silty shales, calcareous shales | Grayish-black, dark gray | Pyrite | Parallel bedding, calcareous nodules, argillaceous nodules | Demirastrites, akidogroptus, pristiograptus | More quantity, less species | 102.00–222.00 | 75.00–152.00 | 4.00–87.40 | 2.40–3.30 |

had deep water when they were deposited and consisted of sediments of deep-water argillaceous continental-shelf facies. The GR curve and the curves of porosity, gas content, and brittle mineral content show an decreasing trend from Long-$I_1$ to Long-$I_2$, and the values of these parameters of Long-$I_1$ are higher than those of Long-$I_2$ as a whole. The values of these parameters are increasingly high in portions closer to the bottom of the Longmaxi Formation, as is their increased amplitude. They show a relatively high peak at the interface between the Longmaxi and Wufeng members and then gradually decrease from the Long-$I_2$ Submember. The sonic interval transit time increases in a bell shape on the interface and then becomes a box shape and varies slightly from the Long-$I_2$ Submember. The DL curve shows an opposite trend compared to the AC curve. Specifically, the density increases in a funnel shape first and then decreases in a thick-box shape from Long-$I_2$. TOC is an important parameter for determining the quality of shale gas reservoirs, and its law of change in the study area agrees well with the GL values, porosity, gas content, and brittle mineral content. The TOC decreases from greater than 2% to less than 2% from the Long-$I_1$ Submember to the Long-$I_2$ Submember at the boundary and decreases upward in a bell shape in the Long-$I_1$ Submember. The porosity shows similar variation characteristics to the sonic interval transit time. It varies in the range of 5–8% in the Long-$I_1$ Submember. Therefore, all parameters indicate that the Long-$I_1$ Submember is the sweet spot in the study area, which has GR values of 200–500 API, average resistivity of approximately 100 $\Omega \cdot$ m, siliceous mineral content of 50%, clay content of 20%, and high pyrite content of approximately 4%. The DL curve is another important basis for determining the Long-$I_1$ Sublayer, which has a density of 2.1–2.5 g/cm$^3$.

The thicknesses of the four sublayers in the Long-$I_1$ Submember show consistent variation trends and increases northward. The thickness of the Long-$I_1^1$ Sublayer is 1.3–2.4 m, with an average of 1.9 m. The thickness of the Long-$I_1^2$ Sublayer is 4.6–11 m, with an average of 8.8 m. The thickness of the Long-$I_1^3$ Sublayer is 3.5–7.5 m, with an average of 5.0 m. The thickness of the Long-$I_1^4$ Sublayer is 6.4–27 m, with an average of 15.7 m.

As shown in the comprehensive logging interpretation profile (Figure 3.10), the quality of the shale gas reservoirs increases in portions closer to the bottom of the Longmaxi Formation. Among the

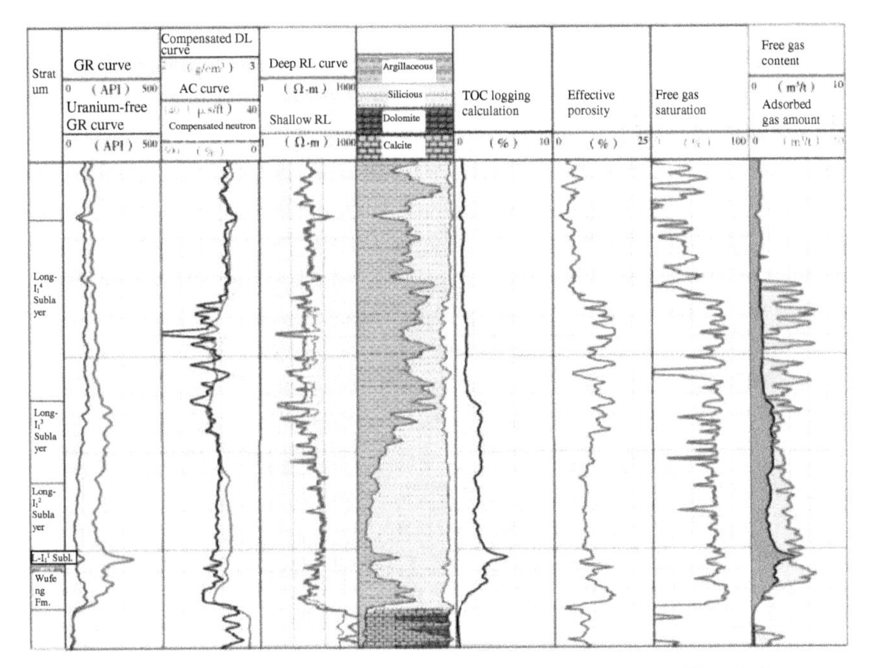


**Figure 3.10.** The comprehensive logging interpretation profile of the target horizons.
*Note*: Subl. refers to Sublayer.

four sublayers of the Long-$I_1$ Submember, the Long-$I_1^1$ and Long-$I_1^3$ Sublayers have higher quality than the Long-$I_1^2$ and Long-$I_1^4$ Sublayers, and the Long-$I_1^1$ Sublayer has the highest quality. All the parameters indicate that the Long-$I_1^1$ Sublayer is the most favorable for shale gas development. It mainly contains Class-I reservoirs and has a large continuous thickness. Although there is a 0.5 m thick interlayer of Class-II reservoirs at the top area, the Long-$I_1^1$ Sublayer is considered having high-quality shales overall. The Long-$I_1^1$ Sublayer has TOC of 6.0–11.0% (average: 8%), porosity of 6–9% (average: 6%), and total gas content of 4–10 $m^3/t$ (average: 7 $m^3/t$). Moreover, the brittleness index of the sublayer was calculated to be 50–65% according to Young's modulus and Poisson's ratio and is generally high. Therefore, the Long-$I_1^1$ Sublayer of the Long-$I_1$ Submember is the optimal sweet spot in this area, and its various indices are the best among the four sublayers.

Figure 3.11 shows the WE-trending well profile for fine-scale correlation of reservoirs in the Long-I$_1$ Submember of the Wufeng Formation in the Changning shale gas field. Among these wells, Wells Ning-203, Ning-201, and Ning-209 are located within the 3D seismic exploration area, with wells Ning-201 and Ning-209 lying in the core of the syncline and well Ning-203, in the northern flank of the syncline. Wells Ning-211 and Ning-212 lie outside the 3D seismic exploration area and are far away from the syncline structure. According to the shale reservoir classification criteria and logging interpretation results, the shale reservoirs of the Wufeng–Longmaxi Formations in the Changning shale gas field are located in the Long-I$_1$ Submember of the Wufeng Formation, with the longitudinal thickness of Class-I and -II reservoirs (Long-I$_1^1$ and Long-I$_1^1$ Sublayers) generally varying in the range of 19.5–35 m.

The well profiles run approximately across the 3D seismic exploration area of the Changning formation from east to west. The green box in the section location map denotes the range of the Changning shale gas field, and the red box in the map is the range of the 3D seismic exploration area. As shown in the well profile, the thickness of the

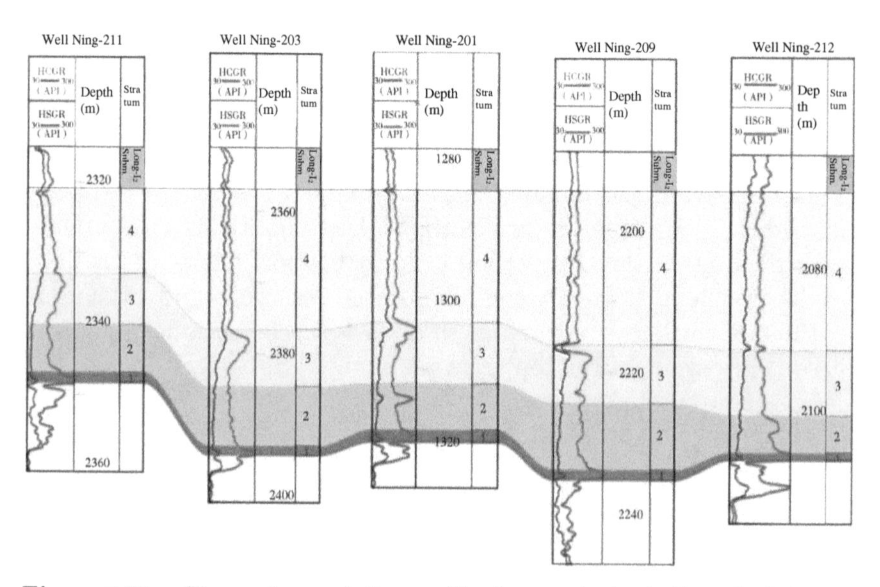


**Figure 3.11.** Fine-scale correlation profile of reservoirs in the Long-I$_1$ Submember of the Wufeng Formation in the Changning shale gas field.
*Note*: Subm. refers to Submember.

Class-I and -II reservoirs in the Long-$I_1$ Submember of the Wufeng Formation has a stable distribution overall. It shows a continuous lateral distribution and tends to thin toward the flanks of the syncline and thicken toward the core of the syncline or the depocenter. The thickness of the Class-I and -II high-quality reservoirs is generally 30–40 m. As shown in Figure 3.11, the high-quality reservoirs have higher continuity and larger continuous thickness in the nucleus of the syncline, while more interlayers occur in areas away from the nucleus of the syncline. These variations are also reflected in the GR curves, including the uranium-free GR curve (high-resolution environment natural computed gamma ray (HCGR)) and the standard GR curve (hostile environment standard gamma ray (HSGR)). As shown in Figure 3.11, the GR curves indicate that the high-quality reservoirs have increasingly poor continuity and decreasing continuous thickness. These changes are noticeable in the areas at the locations of wells Ning-211 and Ning-212.

## 3.3. Seismic petrophysical analysis of shale gas reservoirs

The seismic petrophysical analysis aims to link seismic data and its derived elastic parameters to reservoir evaluation parameters. It is used to study the relationships between reservoir properties and seismic response, especially relationships between the lithological, physical, and hydrocarbon-bearing properties of rocks and their seismic response characteristics, thus guiding the seismic interpretation, identification, and fluid prediction of reservoirs.

### 3.3.1. *Characteristics of elastic parameters of sweet spots*

As the foundation of reservoir prediction, the analysis of the elastic parameters of reservoirs is to make a statistical analysis of log curves and their interpretation results using the method such as the cross plots of parameters and to determine the petrophysical characteristics and sensitive parameters of reservoirs. The raw data used to calculate elastic parameters include DL and AC curves.

In practical production, the shale reservoirs in the study area are generally classified into three classes according to their TOC content.

Class-I shale reservoirs have high quality and TOC >3%, Class-II shale reservoirs have TOC of 2–3%, and Class-III shale reservoirs are common reservoirs with TOC <2%. Figure 3.12 shows the cross plots of common petrophysical elastic parameters of the three classes of reservoirs in the Longmaxi Formation at the locations of three vertical wells in the study area and the cross plots of limestones as surrounding rocks at the bottom of the formation. Density was obtained from logs, P- and S-wave velocities were inferred from the sonic interval transit time, and the ratio of P- to S-wave velocity (the $v_p/v_s$ ratio) and Poisson's ratio were calculated using basic petrophysical formulas. As shown in the cross plots, red dots represent data on Class-I shale reservoirs, yellow dots denote data on Class-II shale reservoirs, green dots represent data on Class-III shale reservoirs, and black dots represent data on the limestones.

The cross plots indicate that: (i) The limestones (black dots) can be effectively distinguished from shale, indicating that the shale

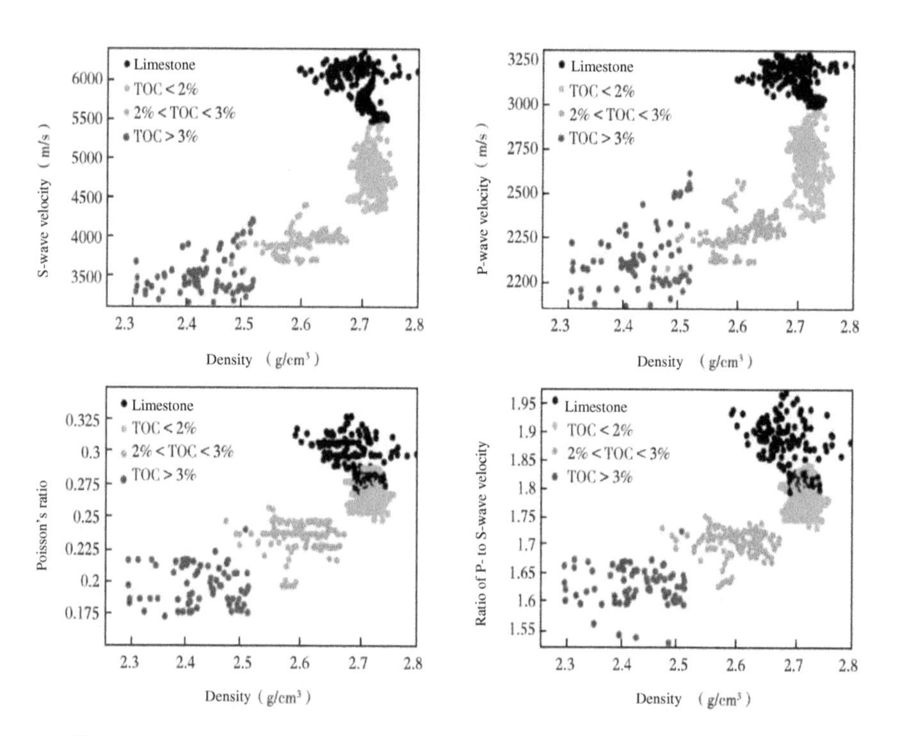


**Figure 3.12.** Cross plots of elastic parameters of shale gas reservoirs.

reservoirs and the surrounding rocks have significantly different characteristics of P- and S-wave velocities, Poisson's ratio, density, and $v_p/v_s$ ratio. The reservoirs in the Longmaxi Formation have significantly low density, low P- and S-wave velocities, low Poisson's ratio, and low $v_p/v_s$ ratio. Moreover, the shales and limestones can be accurately identified using their seismic elastic parameters. (ii) The results of the petrophysical analysis also show that density, P-wave velocity, P-wave impedance, Poisson's ratio, and $v_p/v_s$ ratio significantly decrease with increasing TOC content, providing a good petrophysical foundation for quantitative seismic prediction of Class-I shale reservoirs.

### 3.3.2. *Selecting the optimal elastic parameters sensitive to sweet spots*

To select the optimal elastic parameters sensitive to TOC content, a regression analysis was performed on the TOC content and common elastic parameters obtained from seismic surveys of the target horizons. The regression analysis results are shown in cross plots in Figure 3.13, in which the $y$-axes represent the TOC content obtained from logging interpretation and the $x$-axes represent various elastic parameters such as density and velocity.

As shown in Figure 3.13, there is a significant negative correlation between TOC content and the elastic parameters including density, P- and S-wave velocities, Poisson's ratio, and $v_p/v_s$ ratio. Among them, density is the most sensitive to the TOC content of shale gas reservoirs, with a linear regression correlation coefficient of $-0.8701$. From the perspective of seismic interpretation, density can be obtained through pre-stack inversion, and then the TOC content can be further derived. In sum, the results of the regression analysis can be used as the basis of TOC content prediction from seismic data.

Gas content is another important parameter used to evaluate the potential of shale gas reservoirs. However, since gas occurs in multiple states in shale gas reservoirs, the basic research is still insufficient and geophysical prediction is still in the trial stage. This study also attempted to select the elastic parameters that are sensitive to gas content following the same principles followed in the selection of parameters sensitive to TOC content. The correlation

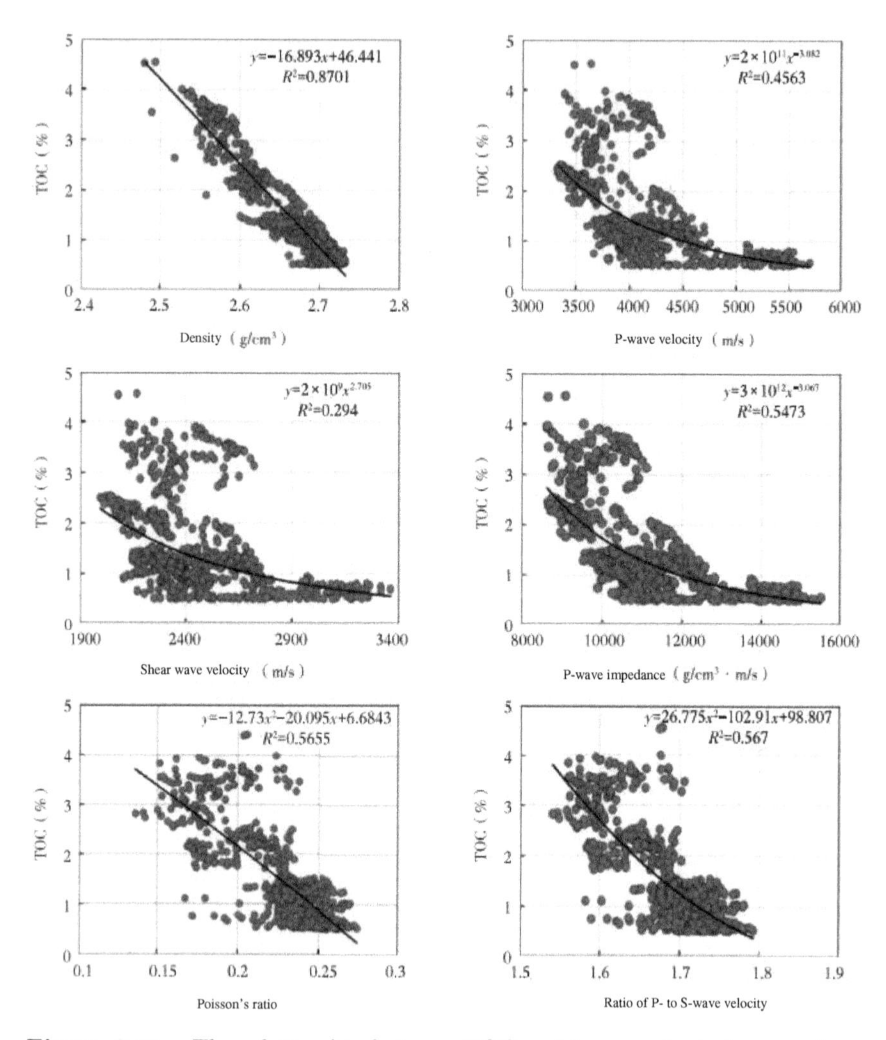


**Figure 3.13.** The relationship between TOC and key geophysical parameters.

coefficients between the gas content and various parameters are shown in Figure 3.14. The parameters that are the most sensitive to gas content include TOC content, density, elastic impedance (EI), P-wave impedance, bulk modulus (K), and Young's modulus (E), of which the absolute values of the correlation coefficients increase.

Many studies have shown that gas content is closely related to porosity and TOC content; the higher the porosity and TOC content,

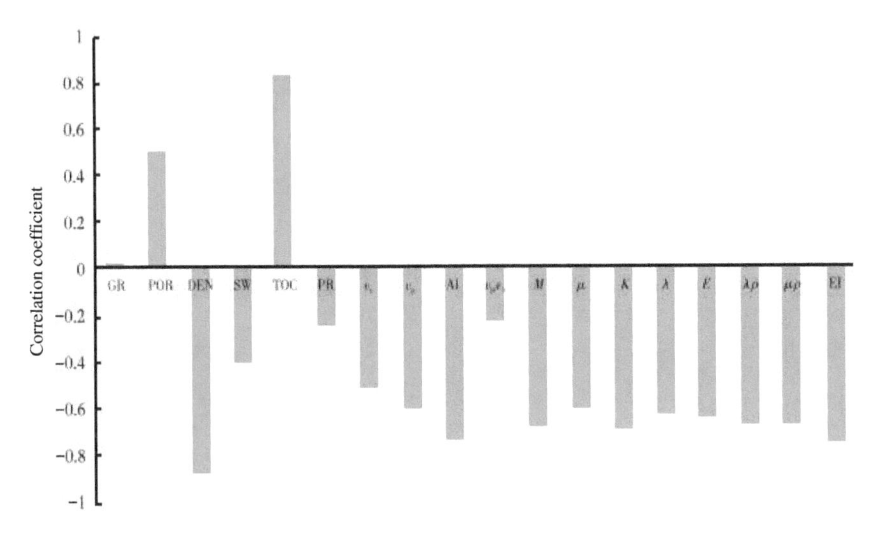


**Figure 3.14.** Correlation between gas content and various elastic parameters.

the higher the gas content and the better the performance of shale gas exploitation. It is generally considered that there is a positive linear correlation between gas content and TOC content and that gas content can be obtained through linear fitting using TOC content.

However, the results of the petrophysical analysis in this study show that the relationship between TOC content and gas content varies with maturity. Figure 3.15(a) shows the relationship between TOC content and total gas content of the Barnett shale gas reservoirs in the Fort Worth Basin, U.S. The Barnett shale gas reservoirs have a relatively low maturity, and there is a good positive linear correlation between their TOC content and gas content, and, thereby, their gas content can be directly derived from the TOC content. In the U.S., shale gas content is mostly predicted using this method. Figure 3.15(b) shows the relationship between the TOC content and gas content of the shale gas reservoirs in the Changning area, the Sichuan Basin. There is a weak positive correlation between the TOC content and gas content. Therefore, it is unsuitable to predict the gas content of the reservoirs in the Changning area using the method of quantitatively predicting high-quality shale gas reservoirs using TOC content. Therefore, a fuzzy optimization technique was applied for the integrated quantitative evaluation of sweet spots in the Changning block.

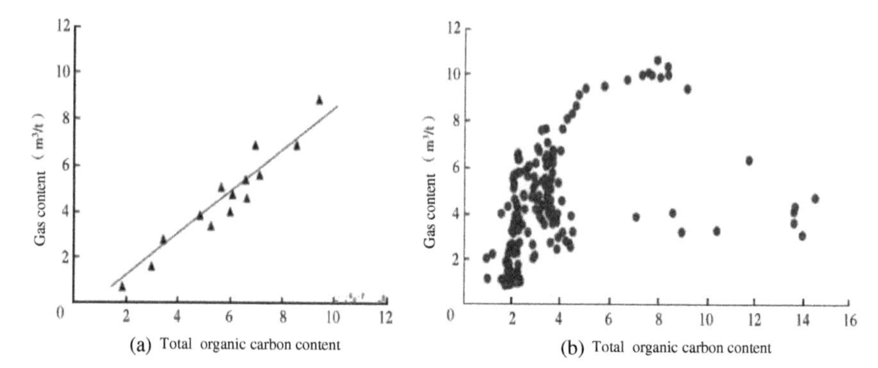


**Figure 3.15.** Relationships between the TOC content and the total gas content of shale gas reservoirs with different maturities. (a) Barnett shale gas reservoirs in the Fort Worth Basin, Fort Worth Basin, U.S. (b) Shale gas reservoirs in the Changning area, Sichuan Basin, China.

Studies have revealed that the main reason for the different relationships is that the geological conditions of shale gas reservoirs in China are fundamentally different from those in North America. The Longmaxi Formation in the study area and the Barnett shale in the Fort Worth Basin were compared as follows. The shale gas reservoirs in the Longmaxi Formation in the Changning area were deposited in the Lower Silurian and have a high maturity, with $R_\mathrm{o}$ greater than 2.0% and of 2.0–3.5% throughout the Sichuan Basin. They have undergone multi-phase tectonism. As a result, some areas with high TOC content have experienced natural gas escape, leaving solid organic matter. Therefore, these areas have no high gas content despite high TOC content. By contrast, some areas with low TOC content have favorable ga-bearing properties owing to good preservation conditions. Compared to the shale gas reservoirs in the Longmaxi Formation in the Changning area, the shale gas reservoirs in North America, represented by Barnett shale in the Fort Worth Basin, have a shorter geological history, a simpler tectonic history, and generally lower thermal maturity. For example, the Barnett shale has $R_o$ of 0.9–1.7%. Moreover, the shale gas reservoirs in North America have moderate thermal evolution and favorable preservation conditions. Therefore, the high TOC content of these reservoirs leads to high gas content, and thus here is a close correlation between the gas content and TOC content.

## 3.4. Petrophysical modeling of shale gas reservoirs

The seismic petrophysical modeling is designed to establish the quantitative relationships between geophysical elastic parameters and key evaluation parameters of subsurface reservoirs (such as physical and hydrocarbon-bearing properties) and thus to determine the change in the seismic response caused by the change in the physical properties and fluids of the reservoirs. The purpose of seismic petrophysical modeling is to achieve quantitative evaluation and prediction of the reservoirs while minimizing the interpretation risks. Seismic petrophysical analysis and modeling bridge geophysics, reservoir engineering, and reservoir geology and are also the physical foundation for the reservoir parameter prediction and reservoir characterization based on seismic data. This is also true for seismic petrophysical analysis and modeling of shale gas reservoirs.

Shale gas reservoirs are characterized by well-developed ultramicropores, fissures, and bedding, complex pore structures, and complex occurrence states of shale gas. So far, there have been no seismic petrophysical modeling methods suitable for shale gas reservoirs. In practical implications, seismic petrophysical modeling methods for conventional reservoirs are inherited or are used after certain modifications or in combination. As a result, it is difficult to provide a scientific basis for the quantitative seismic interpretation of shale gas reservoirs. Therefore, establishing seismic petrophysical modeling methods suitable for shale gas reservoirs is a hot topic at present and is also an important task of this study.

### 3.4.1. *Petrophysical modeling of conventional reservoirs*

There are many petrophysical models of conventional reservoirs currently, such as Gassmann, Biot, BISQ, and Xu-White.

The Gassmann equation (also called the fluid substitution model) is a basic equation in seismic petrophysical analysis. It mainly describes the change in modulus of a rock from a dry state to a saturated state of the rock and is primarily used to calculate the elastic parameters of a rock in the saturated state (e.g., P- and S-wave velocities and density) assuming that the modulus or the P- and S-wave velocities and density of the rock in a dry state are known. The Gassmann equation is derived on the premise that the fluids in the reservoir pores have reached equilibrium, that is, no pressure gradient exists. However, it is only applicable to low-frequency conditions and returns accurate results only when the frequency is low enough. Since seismic waves generally have a main frequency range of 20–60 Hz, it is unsuitable to conduct seismic petrophysical modeling directly using the Gassmann equation.

Based on the acoustic wave equation of porous continuous media, Biot established the relational expression between the parameters of porous reservoir media and the parameters, such as the velocity, frequency, and attenuation of seismic waves, using the method of mechanics of continuous media. The Biot model not only accurately describes the effects of rock matrix properties and pore fluid saturation on the elastic modulus of rocks but also characterizes the interactions between pore fluid and rock matrix, including the viscosity and inertia mechanism of pore fluids during the propagation of seismic waves. Compared with the Gassmann equation, the most prominent advantage of the Biot model is that it is a suitable model to calculate the elastic parameters of porous rocks under any frequencies. However, this model does not consider the jetting effect of fluids under high frequencies. Therefore, if the frequency is very high, the reservoir velocity calculated using this model may not be accurate, but errors can be ignored within the seismic frequency range.

The combined Biot and Squirt (BISQ) model introduces further improvements over the Biot model. It considers the effects of two different flow patterns of pore fluids in reservoirs on the velocity, dispersion, and attenuation of seismic waves. Compared to the Biot model, the BISQ model describes a law that is closer to the propagation law of seismic waves in real reservoir conditions.

While studying sandstone and mudstone reservoirs, Xu-White established the relationships between the velocity of porous media and the parameters of each rock component, such as velocity, content, and gas saturation. The Xu-White model predicts acoustic wave velocity mainly using rock porosity, water saturation, and clay mineral content. In this model, factors affecting the seismic wave velocity are considered to mainly include clay mineral components and their content, reservoir pore pressure, rock cementation, pore morphology, and the plane porosity of reservoirs.

### 3.4.2. *Petrophysical modeling of shale gas reservoirs*

Shale gas has complex occurrence states, including both free gas and adsorbed gas. Absorbed gas accounts for a high proportion of over 30% in the study area and has a significant impact on seismic elastic parameters. Therefore, the adsorbed gas in a shale gas reservoir cannot be ignored as in the seismic petrophysical modeling of conventional reservoirs. Conventional petrophysical modeling techniques like the Gassmann equation cannot be directly used for shale gas reservoirs, and it is extremely difficult to achieve accurate seismic petrophysical simulation. Relevant techniques for the seismic petrophysical modeling of shale gas reservoirs are still under research.

In this study, the self-consistent approximation (SCA) model and the Gassmann fluid substitution model were combined to establish the seismic petrophysical model of shale gas reservoirs. The SCA model was used to build a rock matrix model by processing rock matrix minerals, clay, pores, and fissures. Then the Gassmann model was used to process the fluids. In the process of the modeling, the TOC content was divided into three parts, i.e., solid organic matter, adsorbed gas, and free gas. The solid organic matter and the adsorbed gas were considered integrated parts of the rock matrix and were processed using the SCA model, while the free gas was processed using the fluid substitution model (just like natural gas in conventional reservoirs), thus establishing the seismic petrophysical model of shale reservoirs. The modeling method and flowchart are shown in Figure 3.16. Using this method, the quantitative relationships between the TOC content and the density as well as the seismic wave velocity were determined for the quantitative prediction of TOC content.

### 3.4.3. *Frameworks for quantitative interpretation of reservoirs*

Using the petrophysical reservoir model illustrated in Section 3.4.2, the seismic petrophysical simulation was conducted, and the

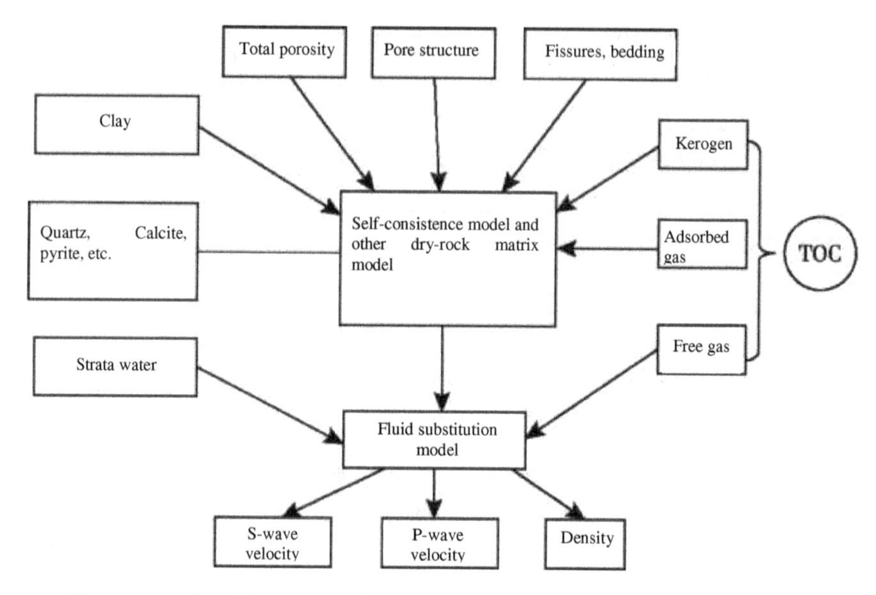


**Figure 3.16.** Flowchart of petrophysical modeling of shale reservoirs.

relationships between seismic petrophysical elastic parameters and key evaluation parameters of shale gas reservoirs were determined. Based on this, the law of change in seismic elastic parameters under different reservoir physical properties and gas-bearing conditions was defined to guide the geophysical prediction of sweet spots. Then, the quantitative relationships between the TOC content and the P- and S-wave velocities, as well as the density under different porosities and free gas content, were defined through a series of simulations, thus forming frameworks for the quantitative prediction and interpretation of the TOC content.

Figure 3.17 shows the quantitative relationship between TOC and P-wave velocity. The reservoirs in the study area have relatively stable water saturation (averaging 30%). Therefore, the water saturation remained unchanged at 30% and the porosity was set to 0–10% in the seismic petrophysical simulation. It is noteworthy that the porosity refers to the total porosity, adequately covering all types of reservoirs in the target horizons. Since the water saturation was set to a constant, the free gas content increased proportionally with an increase

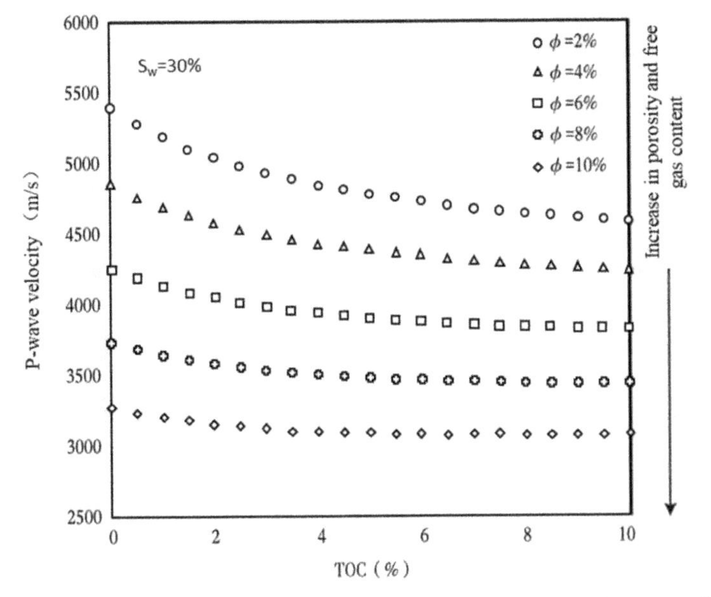


**Figure 3.17.** Quantitative relationship between P-wave velocity and TOC content.

in the porosity. Moreover, the TOC content in this quantitative relationship included solid organic matter, adsorbed gas, and free gas. The y-axis represents P-wave velocity and the x-axis, TOC content in Figure 3.17. As shown in this figure, an increase in the TOC content will inevitably lead to a decrease in P-wave velocity, regardless of porosity. When the TOC content is fixed, the increase in porosity and free gas content will also lead to a decrease in P-wave velocity. However, the influence of the TOC content on P-wave velocity varies under different porosities. For instance, in the case that the TOC content increases from 2% to 10%, the P-wave velocity decreases from 5,000 m/s to 4,500 m/s (by 500 m/s or 10%) when the porosity is 2% but decreases from 3,250 m/s to 3,050 m/s (by 200 m/s or 6%) when the porosity is 10%. This result indicates that the P-wave velocity is more affected by the variation in the amount of organic matter (solid kerogen and free gas) in the rock matrix in the case of low porosity but is more affected by the free gas content in the case of high porosity.

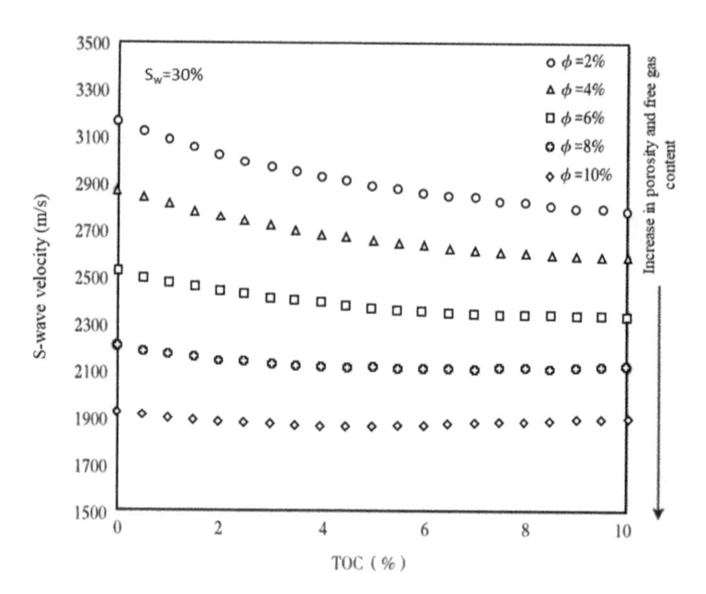


**Figure 3.18.** Quantitative relationship between S-wave velocity and TOC content.

Figure 3.18 shows the quantitative relationship between TOC content and S-wave velocity, with the model parameters being set to the same values as those used in the P-wave velocity simulation. The y-axis represents the S-wave velocity in this figure. The results show that an increase in the TOC content also leads to a decrease in S-wave velocity, regardless of porosity. With the TOC content fixed, an increase in porosity and free gas content will also lead to a decrease in S-wave velocity. The impact of TOC content on S-wave velocity varies with porosity. For instance, in the case that the TOC content rises from 2% to 10%, the S-wave velocity decreases from 3,080 m/s to 2760 m/s (by about 320 m/s or 10%) when the porosity is 2% but remains almost unchanged when the porosity is 10%. This also indicates that the S-wave velocity is more affected by the TOC content in the case of low porosity but is more affected by the free gas content in the case of high porosity.

Figure 3.19 shows the quantitative relationship between TOC content and density, where, the y-axis denotes density. The results show that the density decreases gradually with an increase in the TOC

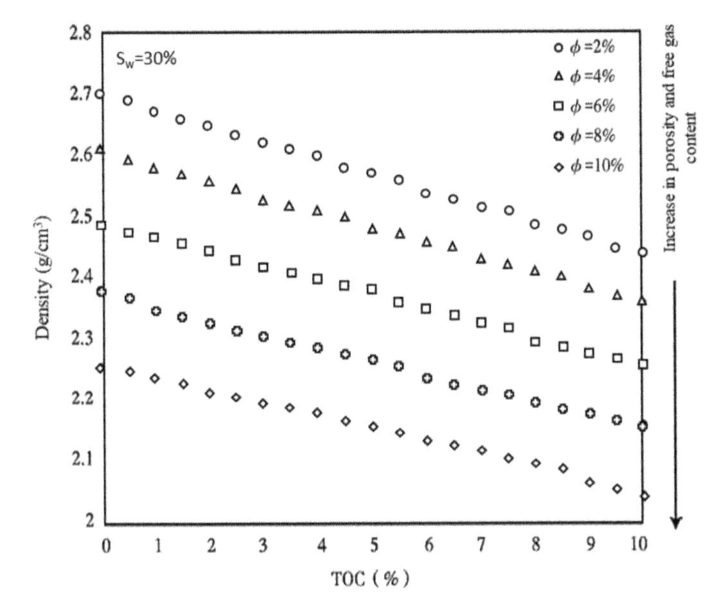


**Figure 3.19.** Quantitative relationship between density and TOC content.

content, regardless of the porosity. Interestingly, as the TOC content varies, the variation trend of density is completely different from that of the P-wave velocity. In particular, there is almost a linear correlation between the density and the change in the TOC content, with each porosity value corresponding to a linear relationship. As clearly shown in Figure 3.19, a change in both TOC content and porosity (free gas) will cause significant variations in density, regardless of the porosity. In the case that the TOC content rises from 2% to 10%, the density decreases from 2.7 g/cm$^3$ to 2.45 g/cm$^3$ (by 0.25 g/cm$^3$ or 10.2%) when the porosity is 2%, and decreases from 2.27 g/cm$^3$ to 2.04 g/cm$^3$ (by 0.23 g/cm$^3$ or 10.1%) when the porosity is 10%. This result indicates that the impact of TOC content on density is roughly the same for different porosity values. This result also reveals that gas content and physical properties also affect the density. Density is the most sensitive elastic parameter of shale gas reservoirs and can reflect the variations in both adsorbed gas and free gas contents.

## 3.5. Seismic response characteristics of shale gas sweet spots

### 3.5.1. *Seismic reflection characteristics of shale gas reservoirs*

The statistical analysis of the petrophysical characteristics of the Wufeng–Longmaxi Formations at the target horizons in the study area reveals that both the overlying limestone of the Liangshan Formation and the underlying limestone of the Baota Formation have high density and high seismic wave velocity. The wave impedance of the limestones is much higher than that of the target horizons, creating significant wave impedance differences, thus forming strong seismic reflections at the top and bottom boundaries of the target horizons.

The above conclusion has been proven by the calibration results of synthetic seismograms. As shown by calibration results of a vertical well (Figure 3.20), high impedance transitions to the low impedance at the top boundary of the Longmaxi Formation, giving rise to strong trough reflections. This finding is completely consistent with the results of the lithological analysis. Moreover, the low impedance of the Longmaxi Formation shifts to the high impedance of the Baota Formation at the bottom boundary of the Longmaxi Formation, creating strong peak reflections. Despite the significantly different wave impedance between the Longmaxi and the Wufeng formations, independent interface reflections cannot be formed between the two formations since the Wufeng Formation has a low thickness of less than 10 m. The bottom boundary of the Wufeng–Longmaxi Formations corresponds to stable, strong peak reflections.

Within the Longmaxi Formation, there also exists significantly different wave impedance between the two submembers in the Long-I Member. The Long-$I_1$ Submember has reservoirs with higher quality and significantly lower wave impedance than the Long-$I_2$ Submember, forming trough reflections at its top boundary (Figure 3.20). Therefore, the reservoirs in the Long-$I_1$ Submember are distributed between a set of relatively stable wave troughs and a set of continuous and stable wave peaks. Despite differences in wave impedance between the sublayers, stable seismic events cannot be formed between the sublayers due to the low thickness of the sublayers,

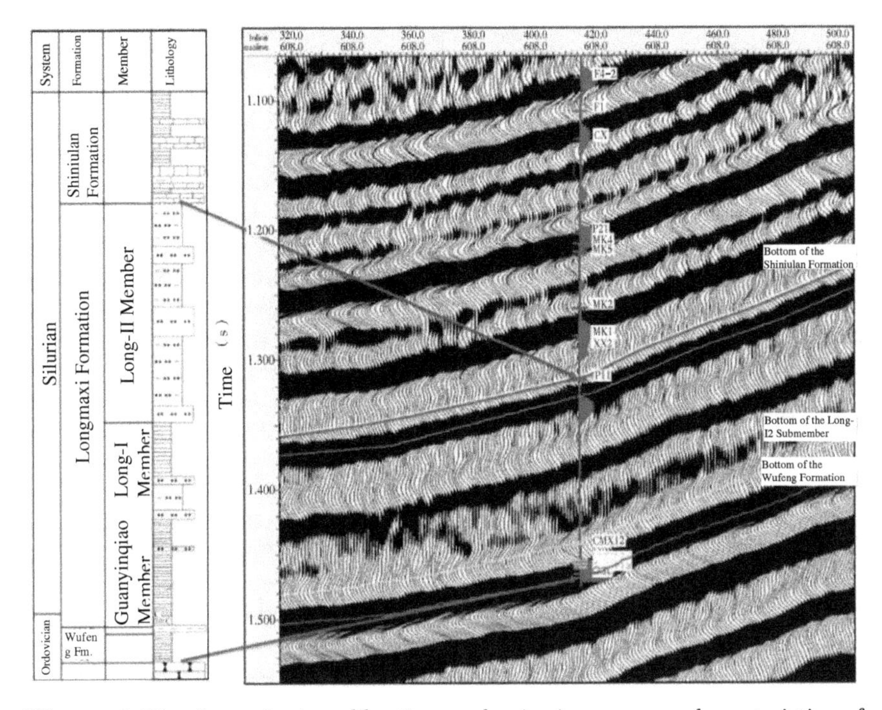


**Figure 3.20.** Log-seismic calibration and seismic response characteristics of reservoirs.

*Note*: Fm. refers to Formation.

making it difficult to distinguish between the reflections among sublayers. High-quality reservoirs (i.e., Long-$I_1^1$, Long-$I_1^2$, and Long-$I_1^3$ Sublayers of the Long-$I_1$ Submember) are distributed in half of a strong wave peak on the seismic profile.

### 3.5.2. *Forward modeling and AVO response of shale gas reservoirs*

Forward modeling of shale gas reservoirs is to: (1) establish a geological model with characteristics approximating those of the actual reservoirs based on parameters such as the P- and S-wave velocities and densities obtained from well logging, drilling, and core measurement; (2) synthesize theoretical seismic profiles according to the seismic reflection theory, and simulate the response characteristics of reservoirs on the seismic profiles under different lithologic, physical-property, and hydrocarbon-bearing conditions; (3) analyze the factors affecting the changes in the seismic response and summarize typical seismic response modes, thus providing a theoretical basis for tracing the reservoir characteristics based on conventional seismic profiles.

The forward modeling of reservoirs is conducted based on the convolution model. When seismic waves cross a formation boundary, they are partially reflected and partially transmitted, and then reflected at various deeper formation boundaries. All the reflected waves formed at various subsurface interfaces are superimposed to form the seismic traces. Mathematically, the forward modeling of reservoirs is a convolution process:

$$S = R * W \tag{3.12}$$

where $S$ refers to seismograms, $R$ refers to the reflection coefficient, and $W$ refers to the wavelet.

The reflection coefficient in the case of normal incidence is

$$R = (p_2v_2 - p_1v_1)/(p_2v_2 + p_1v_1) \tag{3.13}$$

where $\rho_1$, $v_1$, $\rho_2$, and $v_2$ are the densities and velocities of the upper and lower layers, respectively. The model above is the forward model of P-wave impedance under zero offset (self-excited and self-received). Studies have revealed that both high-quality reservoirs and poor reservoirs are impedance interfaces and can form effective seismic reflection. Forward modeling of shale gas reservoirs can be conducted as follows. First, the reflection coefficient sequences at well locations are calculated using both the velocity converted from the sonic interval transit time and the density in the logs. Then, the synthetic seismograms representing the seismic reflection profile under a noiseless condition can be built by convolution of Ricker wavelets of different frequencies with these reflection coefficient sequences. The forward modeling results can be used to analyze the seismic response characteristics of shale reservoirs in order to reduce uncertainties in seismic interpretation. The waveform characteristics of the forward modeling profiles at different frequencies can be analyzed according to the available log data. Then, the waveform characteristics can be summarized to guide the interpretation of the actual forward modeling profiles.

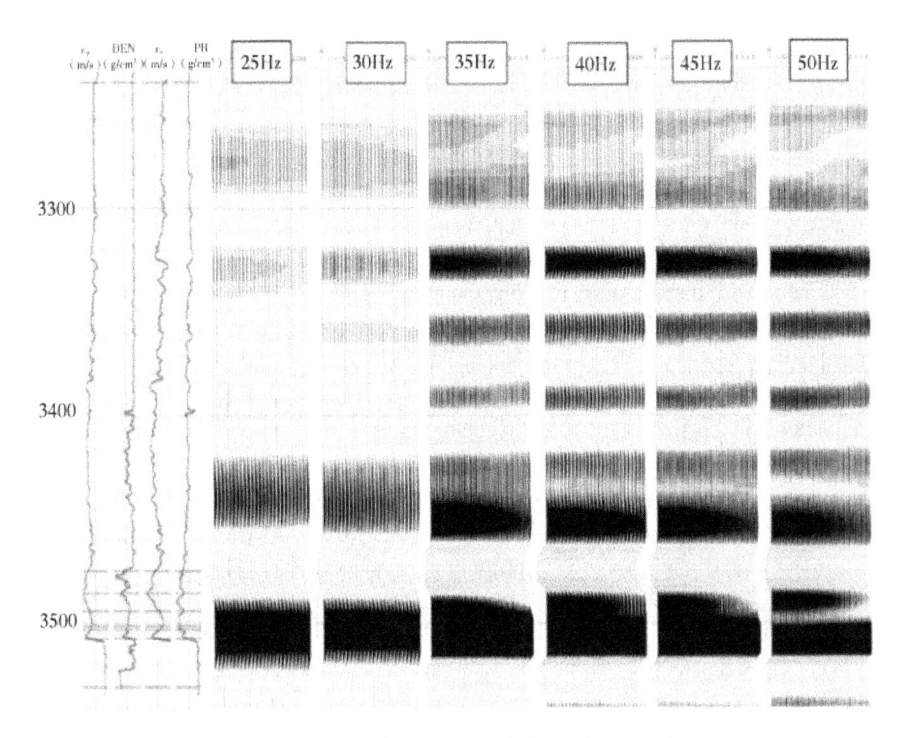


**Figure 3.21.** Forward modeling of reservoirs.

As shown in the forward modeling results, the bottom boundary of the Wufeng Formation is present as a strong peak reflection event, and the top boundary as a relatively strong trough reflection event. These results are consistent with the analysis results in Section 3.5.1.

As shown in Figure 3.21, it is difficult to accurately track the reflections at the top boundaries of high-quality reservoirs (Long-$I_1^1$, Long-$I_1^2$, and Long-$I_1^3$ Sublayers). At low frequencies (25–40 Hz), the boundary appears between the trough at the bottom of the Long-$I_2$ Submember and the peak at the bottom of the Wufeng Formation. At high frequencies (45–60 Hz), the boundary is present as a strong reflection peak between the trough at the bottom of the Long-$I_2$ Submember and the peak at the bottom of the Wufeng Formation. However, due to constraints of the raw data's quality, it is not possible to expand the seismic data beyond the frequency of 45 Hz. Therefore, high-quality reservoirs (Long-$I_1^1$, Long-$I_1^2$, and Long-$I_1^3$ Sublayers) cannot be precisely calibrated through direct interpretation.

Amplitude variation with offset, or amplitude versus offset (AVO) is commonly used in hydrocarbon detection based on seismic data. The AVO technique is mainly designed to investigate the variation in seismic reflection amplitude with offset and predict the variations in lithology and hydrocarbon-bearing property of strata using pre-stack seismic gather data. It is a seismic technique that directly characterizes reservoirs using the information on seismic reflection amplitude.

The AVO forward modeling is very critical in the application of the AVO technique. This AVO forward modeling is mainly to analyze the types and characteristics of AVO anomalies of reservoirs at the location of a known well by substituting the known log data into the AVO forward modeling equation. Then, the physical and hydrocarbon-bearing properties of reservoirs without wells in the same work area can be predicted using the AVO anomaly characteristics. The AVO forward modeling can be used to guide the processing of the AVO data, interpretation, and reservoir calibration, and can also provide a basis for analyzing the law of the seismic response of the physical and hydrocarbon-bearing properties of reservoirs.

As shown in Figure 3.22, when the actual log data were substituted into the forward modeling equation, the seismic amplitude

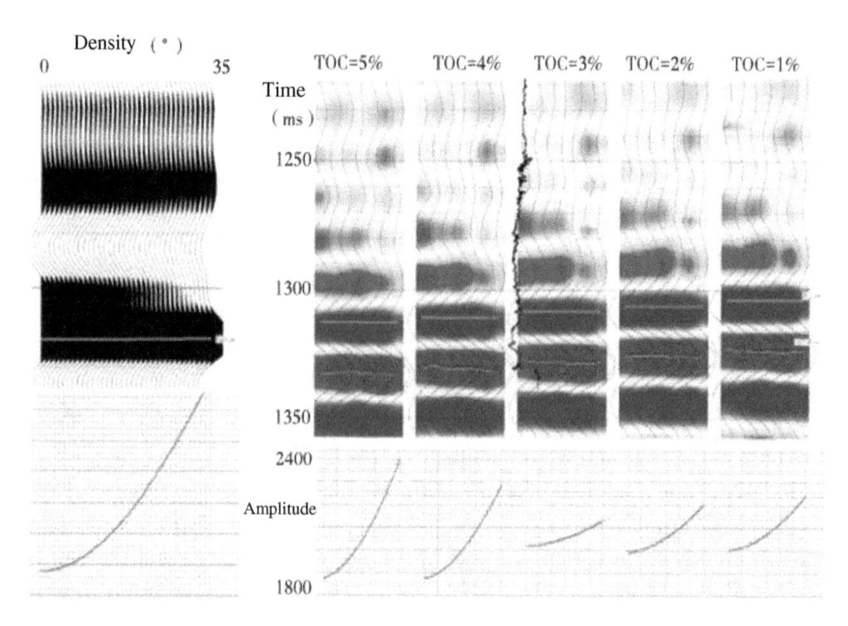


**Figure 3.22.** AVO forward modeling of reservoirs.

showed noticeable Class-III AVO as the offset increased. The AVO response characteristics under different TOC contents were studied by changing fluids and lithology of the reservoirs. As revealed by the results, lower TOC content corresponded to weaker AVO anomalies, and vice versa. Therefore, the AVO anomalies can reflect the changes in the TOC content of the reservoirs and can be utilized as guidance and a basis for reservoir prediction and even the quantitative prediction of the TOC content in the study area.

## 3.6. Summary

(1) The shale gas reservoir intervals of the Longmaxi Formation in the study area have complex lithology, which comprises mudstone, calcareous mudstone, and carbonaceous mudstone besides shale, making it difficult to identify shale gas through logging. The log response of the favorable shale interval in this area is characterized by high natural gamma-ray values, high sonic interval transit time, high neutron values, medium-high PE, low density, and low resistivity. For the image log response, the favorable shale interval in the study area are present as general high resistivity and bright color on the static images and as dark stripes on the dynamic images, indicating beddings are extremely developed in this interval.
(2) According to the national classification criteria for shale gas reservoirs combined with practical production, as well as the data on petrology, sedimentary structures, paleontology, and electrical properties, this study divided the Long-I Member into Long $I_1$ and Long $I_2$ Submembers. Among them, the Long $I_1$ Submember can be further subdivided into four sublayers from bottom to top, namely, Long $I_1^1$, Long-$I_1^2$, Long-$I_1^3$, and Long-$I_1^4$. As indicated by the geological and engineering factors and the log data interpretation results of reservoirs, the Long $I_1$ Submember is superior to the Long $I_2$ Submember in terms of reservoir quality. Among the four sublayers of the Long $I_1$ Submember, Long $I_1^1$ is the most favorable and is the true sweet spot in the study area. The planar distribution trend of high-quality reservoirs shows that Class-I and -II high-quality reservoirs tend to be thicker and more continuous toward the syncline core but tend to be

thinner, less continuous, and more interlayered toward the syncline limbs.

(3) The petrophysical analysis results reveal that the reservoirs in the Longmaxi Formation in the study area are characterized by low density, low P-wave velocity, low Poisson's ratio, and low $v_p/v_s$ ratio. All values of these parameters decrease with an increase in the TOC content. Thus, there is a solid petrophysical foundation for the quantitative prediction of shale reservoirs using seismic techniques.

(4) The sensitivity analysis of the elastic parameters shows that there is a significant negative correlation between the TOC content and the elastic parameters including density, P- and S-wave velocities, Poisson's ratio, and $v_p/v_s$ ratio. Among them, density is the most sensitive to the TOC content of the shale gas reservoirs and is the optimal seismic elastic parameter for TOC content prediction. Moreover, density is also the parameter that is the most sensitive to the variation in gas content. Therefore, density is the most sensitive parameter of shale gas reservoirs in this area.

(5) The petrophysical simulation results of the reservoirs are as follows. The seismic wave velocity is significantly affected by solid TOC content (solid organic matter + adsorbed gas) in the case of low porosity, while the P-wave velocity is mainly influenced by free gas content in the case of high porosity. Density decreases significantly with an increase in porosity, gas content, and solid TOC content (solid organic matter + adsorbed gas), and varies linearly with the TOC content. Regardless of the porosity, the adsorbed gas and free gas contents significantly affect the density.

(6) The top and bottom boundaries of the reservoirs correspond to strong trough reflection and strong peak reflections, respectively, and the bottom boundary of the high-quality reservoirs corresponds to continuously strong peak reflections. At present, the conventional seismic prediction methods can effectively predict the vertical and lateral distribution of Class-I shale reservoirs in the Long$I_1$ Submember and are difficult to distinguish between the sublayers of the Long $I_1$ Submember. The bottom boundary of the Class-I shale reservoir shows noticeable Class-III AVO, which can provide a basis for reservoir prediction and identification.

## Bibliography

Chen S, Zhao WZ, Ouyang YL, *et al.* Comprehensive prediction of shale gas sweet spots based on geophysical properties: A case study of the Lower Silurian Longmaxi Fm in Changning block, Sichuan Basin. *Natural Gas Industry*, 2017, 37 (5): 20–30.

Chen S, Zhao W, Ouyang Y, *et al.* Prediction of sweet spots in shale reservoir based on geophysical well logging and 3D seismic data: A case study of Lower Silurian Longmaxi Formation in W4 block, Sichuan Basin, China. *Energy Exploration & Exploitation*, 2017, 35 (2): 147–171.

Chen ZQ, Yang HF, Wang JB, *et al.* Application of 3D high-precision seismic technology to shale gas exploration: A case study of the large Jiaoshiba shale gas field in the Sichuan Basin. *Natural Gas Industry*, 2016, 36 (2): 9–20.

Chen ZQ. Quantitative seismic prediction technique of marine shale TOC and its application: A case from the Longmaxi Shale Play in the Jiaoshiba area, Sichuan Basin. *Natural Gas Industry*, 2014, 34 (6): 24–29.

Dai JX, Xia XY, Wei YZ. Estimation of natural gas resources and reserves in China. *Oil & Gas Geology*, 2001, 22 (1): 1–8.

Edital Group of Petroleum Geology of Oil and Gas Provinces in Sichuan. *Petroleum Geology of China (Volume 10): Oil and Gas Provinces in Sichuan*. Beijing: Petroleum Industry Press, 1989.

Fatti JL. Detection of gas in sandstone reservoirs using AVO analysis: A 3-D seismic case history using the Geostack technique. *Geophysics*, 1994, 59 (59): 1362–1376.

Gardner GHF, Gardner LW, Gregory AR. Formation velocity and density; the diagnostic basics for stratigraphic traps. *Geophysics*, 1974, 39 (6): 770–780.

Hampson DP, Russell BH, Bankhead B. Simultaneous inversion of pre-stack seismic data. *Seg Technical Program Expanded Abstracts*, 2005, 24 (24): 1633–1642.

He CL, Lv G, Huang TJ, *et al.* Application of pre-stack prediction method of shale gas based on petrophysical analysis in Weiyuan area of the southern Sichuan Basin. *Acta Petrolei Sinica*, 2020, 41 (10): 52–61.

Li WG, Yang SL, Wang ZZ, *et al.* Shale gas development evaluation model based on the fuzzy optimization analysis. *Journal of China Coal Society*, 2013, 38 (2): 264–270.

Li X, Zhou CC, LI CL, *et al.* Advances in petrophysical analysis technology of shale gas. *Well Logging Technology*, 2013, 37 (4): 352–359.

Liu P. *The Rock Physical Property and AVO Forward Modeling of Shale Gas Reservoirs*. Beijing: China University of Geosciences (Beijing), 2012.

Liu SG, Ma WX, Jansa L, *et al.* Characteristics of the shale gas reservoir rocks in the Lower Silurian Longmaxi Formation, East Sichuan basin, China. *Acta Petrologica Sinica*, 2011, 27 (8): 2239–2252.

Pan GT, Xiao QH, Lu SN, *et al.* Subdivision of tectonic units in China. *Geology in China*, 2009, 36 (1): 1–4.

Pu BL, Jiang YL, Wang Y, *et al.* Reservoir-forming conditions and favorable exploration zones of shale gas in Lower Silurian Longmaxi Formation of Sichuan Basin. *Acta Petrolei Sinica*, 2010, 31 (2): 225–230.

Russell B, Dan H, Bankhead B. An inversion primer. *Cseg Recorder*, 2003, 21 (6): 85–93.

Wang ZC, Zhao WZ, Peng HY. Characteristics of multi-source petroleum systems in Sichuan basin. *Petroleum Exploration and Development*, 2002, 29 (2): 26–38.

Yu WH, Zhu W, Bian AF. Role of virtual petrophysical analysis in the prediction of shale gas sweet spots and formation pressure. 2013 Geophysical Exploration Technology Seminar of China Petroleum Society, Hebei, China, July 1, 2013.

Yu X. Physical model of anisotropy of shale gas reservoir. *Petrochemical Industry Technology*, 2016, 23 (11): 166–167.

Zhai GM, Wang JJ. The regularity of oil deposits' distribution. *Acta Petrolei Sinica*, 2000, 21 (1): 1–9.

Zhang JC, Nie HK, Xu B, *et al.* Geological conditions of shale gas accumulation in Sichuan Basin. *Natural Gas Industry*, 2008, 28 (2): 151–116.

Zhang KF, Li CC. Velocity estimation method based on rock-physics model in shale reservoir. *Science Technology and Engineering*, 2019, 19 (11): 10–15.

Zhang TJ, Liu MY, Li DZ, *et al.* Research on shale rock physics modeling. *Offshore Oil*, 2019, 39 (1): 15–20.

Zhao WZ, Wang ZC, Zhang SC, *et al.* Analysis on forming conditions of deep marine reservoirs and their concentration belts in superimposed basins in China. *Chinese Science Bulletin*, 2007, 52 (a1): 9–18.

Zhou MH, Liang QY. Petroleum geological conditions of lower assemblage in Qianzhong uplift and peripheral region. *Marine Origin Petroleum Geology*, 2006, 11 (2): 17–24.

Zhu YP, Liu NR, Martinez A, *et al.* Understanding geophysical responses of shale-gas plays. *Leading Edge*, 2011, 30 (3): 332–338.

https://doi.org/10.1142/9789811283185_0004

# Chapter 4

# Seismic Prediction of Key Parameters for Geological Evaluation of Sweet Spots

Key parameters for geological evaluation, such as total organic carbon (TOC), serve as the material basis for shale gas industrial development. These parameters are mainly obtained by conducting pre-stack seismic inversion and then converting the inversed elastic parameters into geological evaluation parameters according to the petrophysical analysis results. This chapter first introduces the basic principles and characteristics of the simultaneous pre-stack inversion technique. Then, it analyzes the effects of faults on the distribution of sweet spots, determines the criteria for fault classification and interpretation targeting shale gas development, and investigates the effects of the faults of different orders on sweet spot distribution and the gas production from horizontal wells. Finally, this chapter introduces the seismic prediction techniques and methods for various key geological evaluation parameters, such as the porosity, TOC, and gas content of the shale gas reservoirs.

## 4.1. Simultaneous pre-stack inversion

The key evaluation parameters for shale gas reservoirs are all derived using seismic inversion techniques, including the full gather-based simultaneous pre-stack inversion adopted in this study. The simultaneous pre-stack inversion technique, which makes full use of abundant information about the amplitude, frequency, and the amplitude versus offset (AVO) of reflected waves contained in the pre-stack gathers, can accurately determine the P- and S-wave velocities and density, which is the key to the accurate prediction of the evaluation parameters of shale gas reservoirs.

### 4.1.1. *Methods and their principles*

The pre-stack seismic inversion is intended to obtain reliable estimates of the elastic parameters of strata, including P-wave velocity ($v_p$), S-wave velocity ($v_s$), and density ($\rho$), in order to further predict the fluid and lithological characteristics of formations. Numerous geophysicists and geologists have made great efforts in the study of pre-stack seismic inversion. There are many types of pre-stack seismic inversion methods according to algorithms and equations involved, including simultaneous pre-stack inversion.

Compared with post-stack wave impedance inversion, the simultaneous pre-stack inversion features significantly different theoretical basis, data demands, implementation methods, and calculation formulas. Moreover, the post-stack inversion ignores the fact that there inevitably is a close correlation between the P-wave impedance ($Z_P$) and S-wave impedance ($Z_S$) of rocks. In the case of no fluid in reservoirs or no other complex factor, there should be a linear relationship between $v_p$ and $v_s$. Furthermore, the density should be correlated with $v_p$ according to a certain form of the Gardner equation, and many researchers have made great efforts in this correlation.

On the contrary, the simultaneous pre-stack inversion fully considers the coupling relationship between parameters such as P- and S-wave velocities, impedances, and densities, thus increasing the stability and reliability of the inversion algorithm and contributing to solving the strong multiplicity of solutions of inversion. The simultaneous pre-stack inversion is based on the Fatti equation, which is obtained by simplifying the Zoeppritz equation — a basic equation used to describe seismic wave propagation. The algorithm of the simultaneous pre-stack inversion is based on three assumptions: (1) The seismic reflection coefficient is roughly linear. (2) As the functions of angles, the PP- and PS-wave reflection coefficients can be obtained using the Aki–Richards equation and re-derived using equations such as Fatti. (3) There is a linear relationship between the logarithm of P-wave impedance and the S-wave impedance and density as shown in Equations (4.6) and (4.7). Massive experiments have proved that these assumptions apply to most rocks. Based on these three assumptions, the inversion results of real P-wave impedance, S-wave impedance, and density can be obtained by disturbing the initial P-wave impedance model:

$$R_{\mathrm{pp}}(\theta) = c_1 R_p + c_2 R_s + c_3 R_D \tag{4.1}$$

where $c_1 = 1 + tan^2\theta$, $c_2 = -8\gamma^2 tan^2\theta$, $\gamma = V_s/V_p$, $c_3 = -0.5tan^2\theta + 2\gamma^2 sin^2\theta$; and $R_P$, $R_S$, and $R_D$ denote the P-wave reflection coefficient, the S-wave reflection coefficient, and the density, respectively. They can be expressed using Equations (4.2)–(4.4):

$$R_p = \frac{1}{2}\left(\frac{\Delta V_p}{V_p} + \frac{\Delta p}{p}\right) \tag{4.2}$$

$$R_s = \frac{1}{2}\left(\frac{\Delta V_s}{V_p} + \frac{\Delta p}{p}\right) \tag{4.3}$$

$$R_D = \frac{\Delta p}{p} \tag{4.4}$$

The seismic trace $T(\theta)$ of any incident angle in a gather can be expressed using Equation (4.5):

$$T(\theta) = \frac{1}{2} c_1 W(\theta) D L_p - \frac{1}{2} c_1 W(\theta) D L_s - W(\theta) D L_D \tag{4.5}$$

where $L_s = ln(V_s P)$, $L_D = lnp$.

$$lnZ_s = klnZ_p - k_c - \Delta L_s \tag{4.6}$$

$$lnZ_D = mlnZ_p - m_c - \Delta L_D \tag{4.7}$$

Equation (4.5) can be directly used for inversion since there is a correlation between $L_p$ and $L_s$ and between $L_p$ and $L_D$, as shown in Equations (4.6) and (4.7).

Therefore, the simultaneous pre-stack inversion makes full use of the pre-stack information contained in seismic traces with different offsets and well log data (e.g., P-wave, S-wave, and density) for joint inversion, thereby achieving high-precision predictions of the physical properties and hydrocarbon potential of reservoirs. Therefore, this method is widely applied in the semi-quantitative and quantitative characterization of reservoirs. Compared with the post-stack inversion and the traditional pre-stack inversion, the simultaneous pre-stack inversion has significantly improved prediction accuracy.

To obtain high-quality inversion results and improve the stability and reliability of the simultaneous pre-stack inversion, it is necessary to optimize pre-stack gathers. Generally, denoising and moveout correction should be performed in the initial stage. Moreover, the seismic data should be converted into angle gathers since they are offset-domain rather than angle-domain data. Afterward, it is necessary to conduct horizon calibration for each well and seismic data to optimize the time versus depth relationships. Finally, the inversion is carried out after the low-frequency model and the wavelets for inversion are obtained.

### 4.1.2. *Technical advantages and limitations*

The traditional pre-stack inversion, which is mainly based on partial stacks, has been well applied in conventional natural gas exploration due to its strong antinoise ability and high stability. However, the loss of AVO information in the original pre-stack seismic gathers during the partial-stack inversion inevitably poses adverse impacts on the inversion accuracy of elastic parameters. By contrast, the simultaneous pre-stack inversion, which is based on full pre-stack common-reflection-point (CRP) gathers, can improve the inversion accuracy of elastic parameters by fully utilizing the complete AVO information of the original gathers, thus providing a better data basis for sweet spot prediction. The comparison of the inversion processes between the traditional pre-stack inversion and the simultaneous pre-stack inversion used for this study is shown in Figure 4.1.

### 4.1.3. *Gather optimization and its effects*

To improve the noise reduction ability of simultaneous pre-stack inversion and to ensure the inversion accuracy, this study optimized original gathers as follows. First, the original CRP gathers (Figure 4.2(a)) were processed using the residual moveout correction (Figure 4.2(b)) and super gathers (Figure 4.2(c)) to obtain

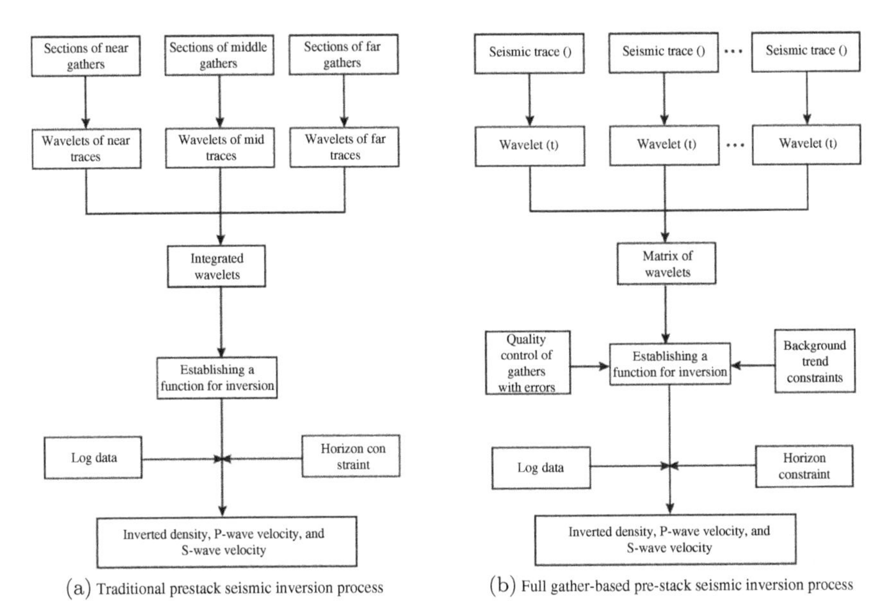


**Figure 4.1.** Comparison of pre-stack inversion processes (modified after Su *et al.*, 2016).

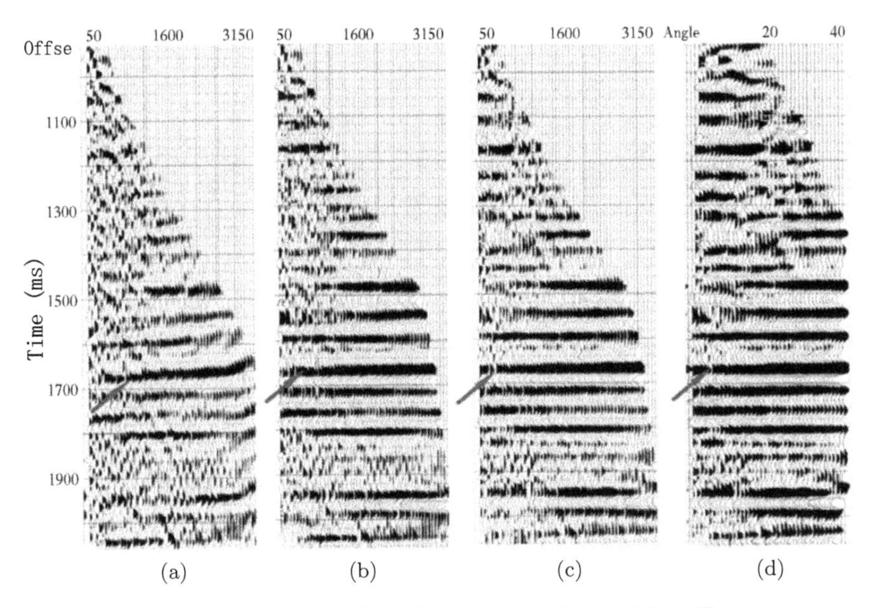


**Figure 4.2.** Pre-stack gather optimization and the effects.

high-quality pre-stack gathers. Then, the CRP gathers were converted into the angle-domain gathers (Figure 4.2(d)) using the interval velocity to facilitate simultaneous pre-stack inversion.

Compared with the original gathers, the optimized gathers have much higher quality, their residual moveout was effectively corrected, and their signal-to-noise ratio was significantly improved. The red arrow in the figure indicates the reflection event corresponding to the target strata. The optimized gathers can be directly used for simultaneous pre-stack inversion.

The inversion accuracy of elastic parameters directly determines the reliability of the prediction results. Therefore, tough quality control measures should be taken during the simultaneous pre-stack inversion. Before the inversion, multiple experiments of parameters were carried out on the entire data volume, with each experiment yielding a set of inversion results. Then, the forward calculation was carried out once based on the inversion results, producing a synthetic gather. Subtracting the original gather from the synthetic gather yielded the error gather. The inversion of the entire data volume cannot be conducted until the error gather approximates to white noise. Figure 4.3 shows the inversion results of the study area after

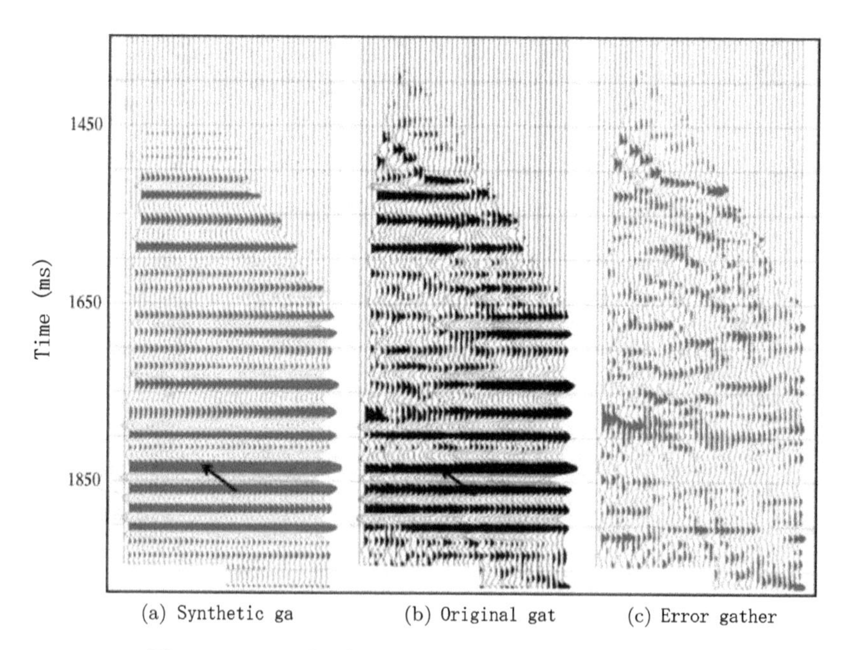


**Figure 4.3.** Quality control of pre-stack inversion.

the quality control, with the synthetic gather, the original gather, and the error gather represented in Figures 4.3(a)–(c), respectively. The error gather was the difference between the original gather and the synthetic gather. As shown in the comparison of these results, the synthetic gather and the original gather tended to be consistent in terms of amplitude, frequency, and AVO characteristics, and the error gather was close to white noise.

After the gather optimization, the full gather-based simultaneous pre-stack inversion was applied to predict the elastic parameters of shale gas reservoirs in the study area, obtaining accurate results. Figure 4.4 shows the comparison of the P-wave impedance obtained through full gather-based simultaneous pre-stack inversion

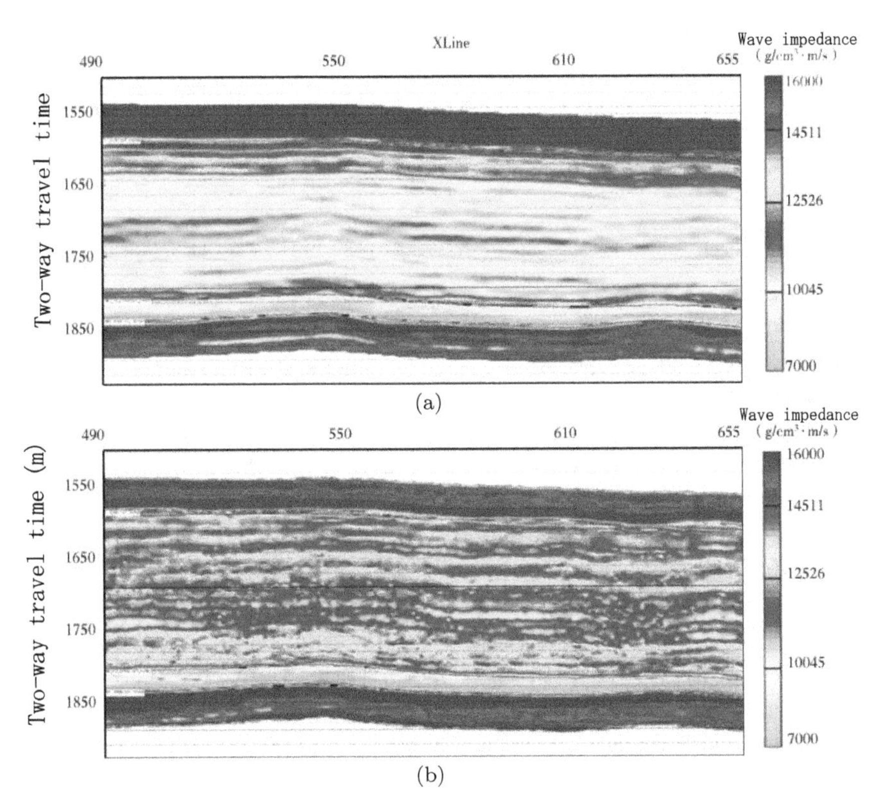


**Figure 4.4.** Comparison of the P-wave impedance obtained from the traditional pre-stack inversion and from the full gather-based simultaneous pre-stack inversion. (a) Pre-stack seismic inversion based on partial stack seismic data. (b) Full gather-based pre-stack simultaneous inversion.

and traditional pre-stack inversion techniques. Figure 4.4(a) shows the P-wave impedance obtained from the inversion based on the partial stack seismic data (traditional pre-stack inversion) and Figure 4.4(b) is the wave impedance obtained from the simultaneous pre-stack inversion based on full gathers. The comparison indicates that the P-wave impedance obtained from full gather-based simultaneous pre-stack inversion included clearer details, was less dependent on the initial model, and can more effectively reflect the variations in subsurface reservoirs. The inversion results indicate that the wave impedance of the target strata is significantly low at a two-way travel time of approximately 1,850 ms.

## 4.2. The classification and interpretation of faults and the effects of faults on sweet spots

### 4.2.1. *Establishing the criteria for the classification and interpretation of faults*

The interpretation and fine-scale characterization of faults are critical to shale gas development. In the early stages of shale gas exploration and exploitation, the seismic data were mainly used for fault interpretation to avoid large faults when drilling horizontal wells. On the one hand, large faults may connect the aquifers above and shale reservoirs below, which may cause water flooding of gas-bearing reservoirs. On the other hand, faults are important channels for oil and gas migration and may facilitate or restrain the flow of gas. The negative or positive effects of faults depends on the size of faults and on the permeability of the surrounding rocks of shale. Areas with faults are generally not conducive to the preservation of shale gas.

Fractures have a great impact on the preservation of shale gas. For thermogenic shale gas reservoirs, gas migration and accumulation are mainly achieved through microfractures but are destroyed by faults and macroscopic cracks. Therefore, strong tectonic movements are inconducive to the preservation of thermogenic shale gas reservoirs. By contrast, biogenic gas reservoirs were formed under active exchange in fresh water. In this case, fractures serve as both the flow pathways for formation water and the migration and accumulation pathways for shale gas, and thus tectonic movements have positive effects. Studies have shown that the Mesozoic–Paleozoic shale gas reservoirs in South China are thermogenic, and thus their preservation conditions are mainly related to fault systems. Similar to uplift and denudation, the faulting tectonism mainly destroys the continuity of cap rocks. The faulting mainly affects the preservation conditions of shale gas reservoirs in South China by affecting the intensity and nature of movements. By contrast, the faulting-caused destruction of cap rocks and the sealing properties of faults directly impact the preservation conditions of the reservoirs.

Massive domestic studies have shown that fault systems significantly influence the formation and preservation of shale gas reservoirs. Large faults may damage gas reservoirs, minor faults may improve the reservoirs, and microfractures serve as reservoir spaces. Therefore, it is necessary to conduct classification and interpretation of faults in complex tectonic areas and to predict the strong tectonic reformation areas and the areas with fractures. The faults in this study were classified as follows: (1) a first-order fault refers to a regional fault that controls regional structures; (2) a second-order fault shows that seismic events are dislocated by more than one seismic event on the seismic profile and has very large throw, a planar length of more than 10 km, and noticeable coherence or curvature response; (3) a third-order fault shows significantly staggered events on the seismic profile and has obvious throw, a planar length of 3–10 km, and relatively noticeable coherence or curvature response; (4) a fourth-order fault has obvious event deflection on the seismic profile and has obvious throw, a planar length of less than 3 km, and some response of the coherence or the curvature attributes (Figure 4.5).

In tectonically active areas with active faults, the formed oil and gas may disperse along faults and fracture networks, and the residual gas should be dominated by adsorbed gas. Moreover, the formation pressure generally decreases in these areas, making hydrocarbon development more difficult. Therefore, in the process of the exploration and exploitation of shale gas in tectonically active areas, it is necessary to accurately locate the fault network and especially stable blocks without large faults. Faults except first-order ones are well developed at the location of well Lu-203 in the Luzhou block. Moreover, this area shows a noticeable seismic response. Therefore, this study researched and established the criteria for the classification

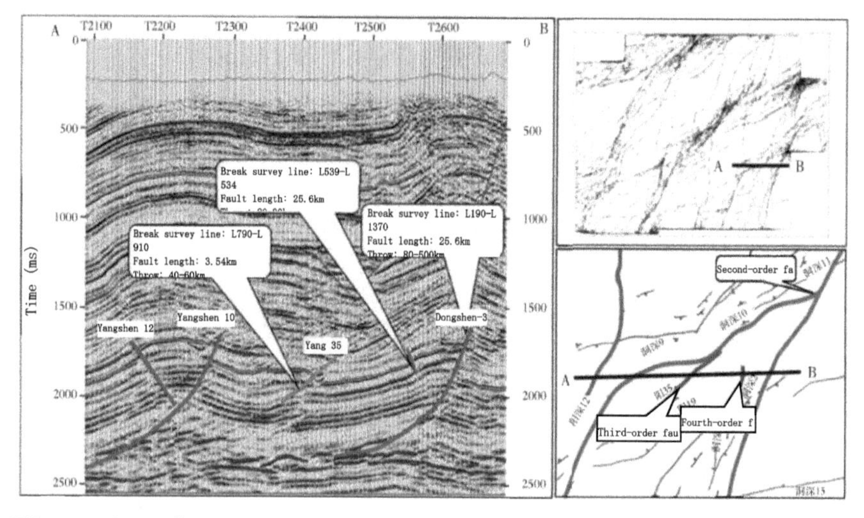


**Figure 4.5.** Seismic response characteristics of faults of different orders in southern Sichuan.

and interpretation of faults by taking the area at the location of well Lu-203 as an example. As shown in Figure 4.6, this area has 38 second-, 116 third-, and 254 fourth-order faults in total according to the interpretation results of faults. The faults in this block consist of two groups, namely, the early northeast- and nearly-east-west-trending faults and the late northeast- and north northeast-trending faults. According to the experience of the U.S., the shale gas exploitation in this area should be carried out at least 80–100 m away from these faults.

In the exploitation of marine shale gas in southern Sichuan, the criteria for the classification and interpretation of faults for shale gas exploration and development are mainly based on seismic characterization combined with geological analysis, as shown in Table 4.1. The bases for the criteria mainly include the scale (length), throw, and influences on shale gas production of the faults.

### 4.2.2. *Influences of faults of different orders on sweet spots*

The Longmaxi Formation in southern Sichuan has favorable caprock, roof, and footwall conditions. Therefore, the most important elements used to characterize the preservation conditions of the

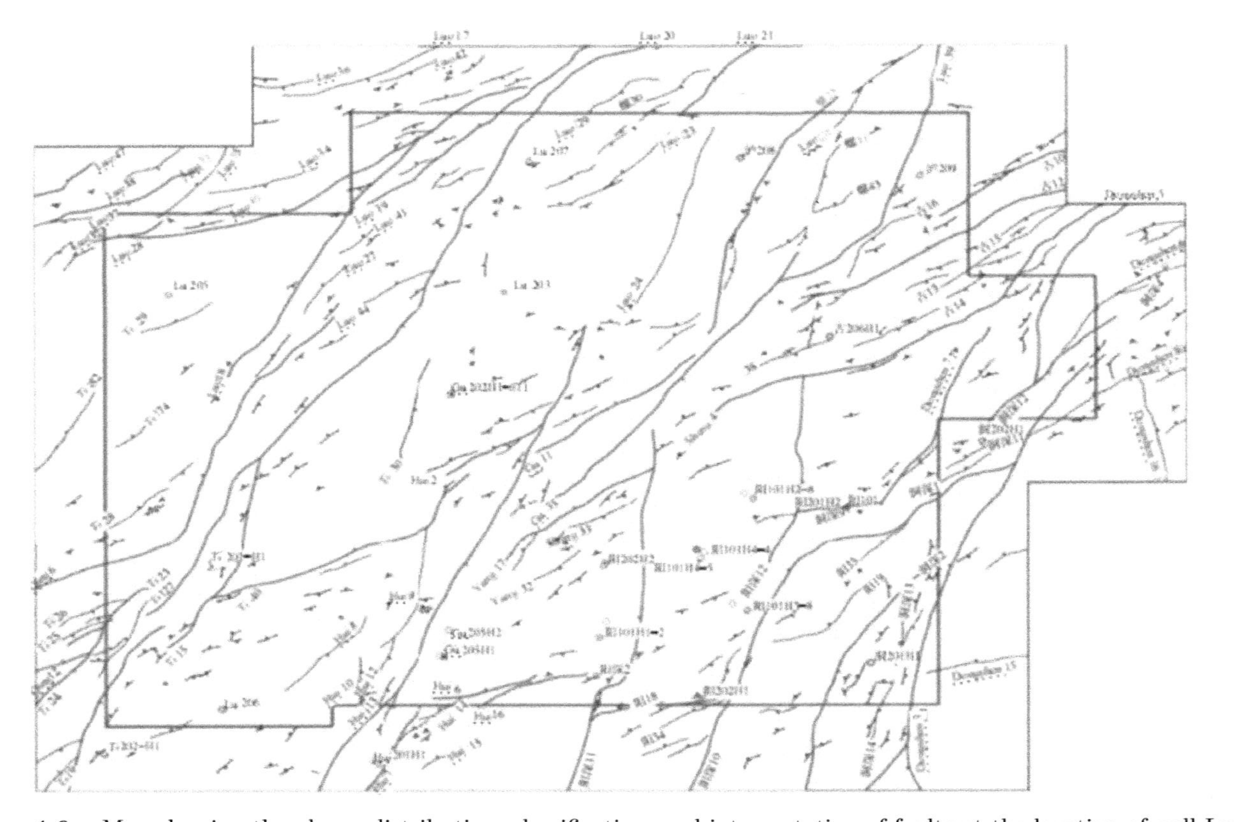

**Figure 4.6.** Map showing the planar distribution, classification, and interpretation of faults at the location of well Lu-203 in the Luzhou block.

**Table 4.1.** Classification criteria and production response measures for shale gas faults in southern Sichuan.

| Fault | Fault throw | Fault length | Effects on shale gas production | Countermeasures |
|---|---|---|---|---|
| First-order fault | >300 m | >15 km | Being highly destructive and not conducive to the preservation of shale gas | Keeping horizontal wells away from such faults |
| Second-order fault | 100–300 m | 10–15 km | Affecting the preservation of shale gas and the shale gas production of wells nearby | Keeping horizontal wells more than 700 m away from such faults |
| Third-order fault | 40–100 m | 3–10 km | Roughly not affecting the preservation of shale gas but may lead to the trajectory out of the compartment to a certain extent | Trying to make the trajectory of horizontal wells avoid crossing such faults along |
| Fourth-order faults | 20–40 m | <3 km | Not affecting the preservation of shale gas but may lead to the trajectory out of the compartment, resulting in invalid footage | Strengthening the geosteering |

Longmaxi Formation include the development degree of faults and the distance from the denudation line. There are many exploitation wells in the Changning block. After classification and interpretation of the faults in the block according to the abovementioned criteria, the impacts of the faults were systematically analyzed according to the shale gas production of horizontal wells, yielding the following findings. (1) The first-order faults (throw greater than 300 m) have a great impact on the shale gas production of the Longmaxi Formation.

The shale gas production in the production test was low within 1.5 km from the first-order faults. For instance, well N7, which is 800 m away from first-order faults, has a pressure coefficient of 1.25 and shale gas production of $11 \times 10^4$ $m^3$/d in the production test. (2) The second-order and third-order faults have little impact on shale gas production. The horizontal wells nearby had very high shale gas production during the production test, with an average greater than $11 \times 10^4$ $m^3$/d (Figure 4.7). The preservation conditions of shale gas at the locations of wells near denudation lines have been compromised. For example, well N8, which is 2,800 m away from the denudation line, has a pressure coefficient of 0.50, and a trace amount of gas has been detected at the location of this well. Another example is well WD1, which is 6,000 m away from the denudation line. This well has a pressure coefficient of 0.92, and a trace amount of gas has also been detected at the location of this well. Moreover, the regions close to the first-order faults, the denudation

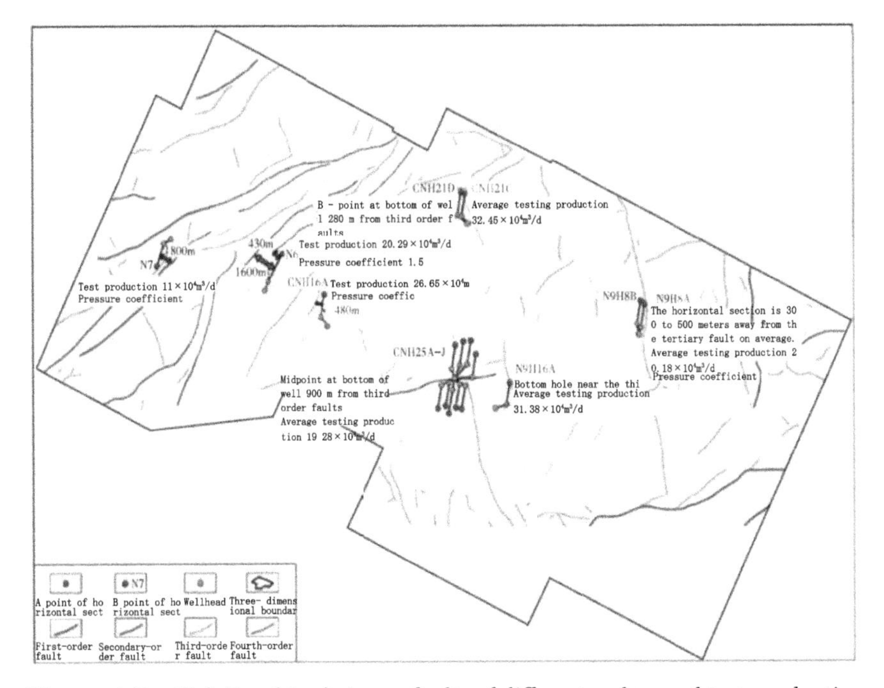


**Figure 4.7.** Relationships between faults of different orders and test production in the Changning area (after Ma *et al.*, 2019).

zone in the Changning block, and the denudation zone of the Leshan-Longnvsi paleo-uplift are characterized by low pressure coefficients. Given this, the pressure coefficient that characterizes the energy and closure degree of underground fluids can be used as a comprehensive parameter to indicate the preservation conditions of the Longmaxi Formation in southern Sichuan. In addition, the pressure coefficient of the Longmaxi Formation increases with increasing burial depth and there is a significant positive correlation between them from the denudation lines in the Changning block and the Leshan-Longnnvsi paleo-uplift toward the Luzhou area.

## 4.3. Seismic prediction of the porosity of shale gas reservoirs

With the progress in the exploration and development of shale gas, the accumulation processes of shale gas and the controlling factors in the high productivity and preservation of shale gas have been better understood. In particular, porosity has been pinpointed as the key factor in determining the gas potential of shale reservoirs. Therefore, the quantitative characterization of shale porosity is both the prerequisite for porosity prediction using seismic data and the basis for shale gas evaluation.

The porosity prediction methods include experimental determination, log interpretation, and combined log-seismic calculation. The experimental determination has high accuracy but is constrained by the availability of core samples and is only applicable in areas where wells have been drilled and cored. The log interpretation also has high accuracy but is restricted to well locations and it is difficult to extrapolate to areas without wells. Therefore, it is necessary to conduct quantitative characterization of porosity using geophysical methods.

Gassmann (1951) proposed the propagation theory of seismic waves within fluid saturated rocks, i.e., the Gassmann equation, by assuming that the porous media are homogeneous and isotropic, all pores are interconnected and filled with frictionless fluids, the fluids in porous media do not interact with pore particles, and the pores in rocks are closed and water does not drain from the pores. The Gassmann equation provided a theoretical basis for the inversion of the elastic wave velocity in porous rocks and the physical properties of the rocks. Biot (1956, 1962) believed that the movement of fluids in the rock pores would cause a loss of seismic wave energy due to friction. Therefore, he studied the influence of both the coefficient of viscosity of fluids and the permeability on elastic waves in two-phase media based on the Gassmann equation. Wyllie *et al.* (1956) proposed an empirical relationship between seismic wave velocity and porosity based on the variation in seismic wave velocity with porosity, i.e., the Wyllie time-average equation, and predicted reservoir porosity using this equation. Researchers such as Raymer *et al.* (1980), Castagna *et al.* (1985), and Han *et al.* (1986) considered the effects of shale content on seismic wave velocity based on the time-average equation proposed by Wyllie. Mavko and Mukerji (1995) and Mavko *et al.* (2003) believed that the poroelastic modulus of dry rocks was closely related to external pressure and porosity when the pressure of pore fluids remained unchanged and calculated the bulk modulus of dry rocks using the elastic modulus and the porosity of both rock matrix and saturated rocks. Quirein *et al.* (2001) calculated the porosity using a regression algorithm according to the relationship between physical properties and porosity of rocks. Russell *et al.* (2003) and Cadeton (2007) analyzed the relationship between seismic attributes and porosity based on both log data and core analysis results and then calculated reservoir porosity using the relationship.

In recent years, many researchers in China have conducted feasibility studies and demonstrations on seismic prediction of gas content in shale reservoirs and carried out experiments using practical data, achieving encouraging results. Zhu GS (1992) proposed an empirical formula for expressing porosity using shale content based on the Wyllie time-average equation, then eliminated the influence of shale using the double-wave method, and finally established a porosity inversion formula using the P- and S-wave velocities. Zhang (1994) studied the seismic stress–strain data based on the Biot theory, established a mathematical model for calculating the elastic parameters of rock strata using seismic data, and explored a new method for the inversion of porosity using seismic data. Sun (2000, 2004) proposed a two-parameter model of porosity and pore structure (known as the frame flexibility coefficient) to describe the elastic wave velocity under the known saturating fluid types and reservoir lithology and determined effects of the pore structure on the elasticity of rocks based on the pore structure type. Yun *et al.* (2001) analyzed the surveyed data of the rock samples and proposed using porosity, temperature, and effective pressure of sandstones to represent their elastic modulus. Shi *et al.* (2010) predicted porosity through the influence of rock density on the seismic wave velocity caused by changes in porosity, pore type, and the compression coefficient of rock matrix. Using the Gassmann equation, He (2012) proposed a porosity inversion method based on the rock matrix model. Zhang *et al.* (2012) proposed a method of obtaining the critical porosity of rocks by inverting the P-wave and S-wave velocities. Wang *et al.* (2013) proposed using the elastic parameters of rocks obtained through seismic elastic inversion combined with the petrophysical analysis to predict porosity using artificial neural networks (ANNs). Zhou *et al.* (2015) calculated the seismic parameters of reservoirs based on the pre-stack inversion method and inverted the reservoir porosity using probabilistic neural networks (PNNs), overcoming the multiplicity of solutions caused by the post-stack wave impedance inversion for reservoir prediction. Zeng *et al.* (2018) calculated the shale porosity of the Longmaxi Formation in the Weiyuan area, Sichuan Basin, based on the linear relationship between P-wave impedance and porosity. Chen *et al.* (2018) integrated log curves and seismic data to carry out a quantitative prediction of fracture porosity through the application of multi-attribute neural networks. Zhang *et al.* (2019) effectively predicted the porosity distribution in the Nanchuan region using the pre-stack inversion and the PNN techniques.

The methods of combined log-seismic calculation for porosity evaluation can be divided into two categories. The first category is based on linear methods: (1) The Wyllie time-average equation is directly used for calculation. This method is more practical for the areas where the lithology is relatively simple and varies slightly in the transverse direction but is not suitable for the areas where reservoir pore types are complex and vary greatly in the transverse direction; (2) The Biot-Gassmann theory-based porosity prediction using seismic data. This method requires many input parameters that are difficult to obtain. (3) Another method is to obtain the relationships between porosity and elastic parameters using the linear or nonlinear statistical analysis of the elastic parameters from log curves and porosity and then calculate the porosity according to the elastic parameters derived from seismic inversion. The second category is based on non linear methods including several porosity prediction methods based on a single or multiple seismic attributes, such as multiple linear regression, stochastic simulation, and neural networks. These methods establish the relationships between seismic attributes and porosity based on the log data or core analysis results. They heavily rely on seismic attributes and lack a theoretical basis since they rarely discuss the internal factors that affect porosity in theory.

### 4.3.1. *Elastic parameters sensitive to the porosity of shale gas reservoirs*

Based on clear logging response characteristics, this study carried out a petrophysical analysis using seismic data as follows. First, the elastic parameters sensitive to the porosity of shale gas reservoirs were determined, and the quantitative relationships between the reservoir evaluation parameters and elastic parameters were established. Then, reliable elastic data volumes were obtained through the high-precision simultaneous pre-stack seismic inversion. Finally, the 3D distribution of porosity was quantitatively predicted using the quantitative relationships between the elastic parameters and porosity. Moreover, the prediction results were comprehensively analyzed and evaluated, thus determining the planar distribution of high-porosity shales.

The analysis of the log curves of well Ning-203, located in the study area (Figure 4.8), shows that there are specific differences between the high-porosity shale reservoirs and the surrounding rocks in terms of elastic parameters such as P-wave velocity, wave impedance, Poisson's ratio, density, and velocity ratio. The reservoirs of the Longmaxi Formation have noticeable characteristics such as low density, low velocity, low Poisson's ratio, low velocity ratio, high gamma ray value, and high Young's modulus — favorable features for the presence of sweet spots. The surveyed porosity and gas saturation curves show that both the Wufeng Formation and the Long-$I_1$ Submember have high porosity and high gas content, indicating that these target strata have a good material foundation and favorable physical properties, and thus have great resource potential.

Based on the petrophysical analysis of the Wufeng Formation and the Long-I Member at the location of well Ning-203, the elastic

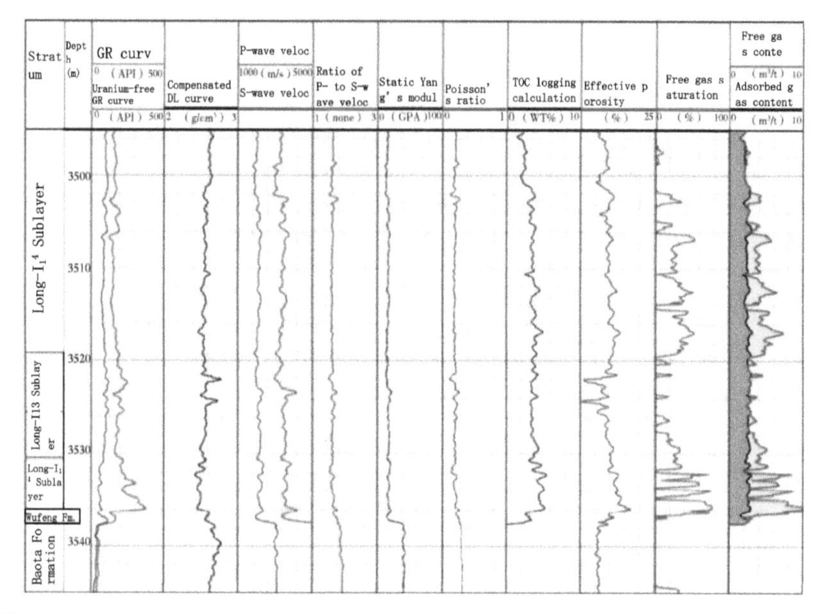


**Figure 4.8.** Log curves of the Wufeng Formation and the Long-I$_1$ Submember at the location of well Ning-203.

*Note*: Fm. refers to Formation.

parameters most sensitive to porosity were selected to establish the quantitative relationship between the elastic parameter and porosity. The cross plots between porosity and density, P-wave velocity, the $v_p/v_s$ ratio, and gamma-ray values (Figure 4.9) show a significant negative correlation between porosity and P-wave velocity, with a correlation coefficient of −0.8308. In other words, higher porosity corresponds to the lower P-wave velocity. This is due to the inherent characteristics of seismic wave propagation. When seismic waves propagate in fluid-filled porous media, significant velocity dispersion and attenuation will occur, and the P-wave velocity will decrease accordingly. These results of the petrophysical analysis are the foundation for the subsequent, more detailed seismic prediction of porosity.

### 4.3.2. *Seismic prediction of porosity*

Porosity is one of the key parameters that determine the characteristics of hydrocarbon reservoirs, and it also directly affects the free gas

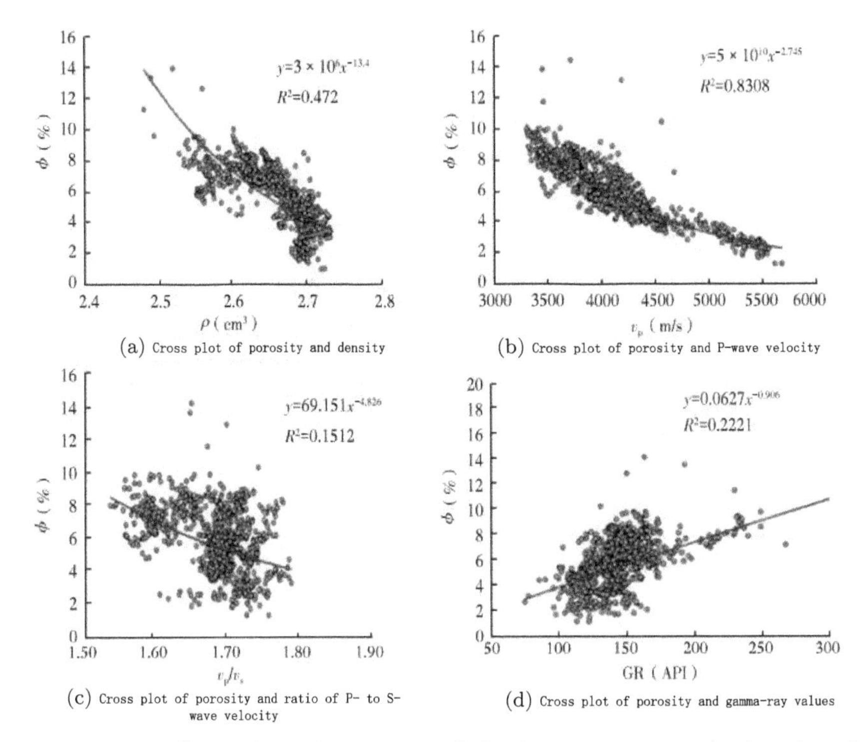


(a) Cross plot of porosity and density

(b) Cross plot of porosity and P-wave velocity

(c) Cross plot of porosity and ratio of P- to S-wave velocity

(d) Cross plot of porosity and gamma-ray values

**Figure 4.9.** Cross plots of porosity and elastic parameters at the location of well Ning-209.

and total gas contents of shale gas reservoirs. Generally speaking, higher porosity corresponds to higher free gas and total gas contents. Currently, the commonly used porosity prediction methods at home and abroad include the Wyllie equation, the Pickett equation, the S-wave equation, and the generalized Cokriging. Most of these methods can only determine porosity at well locations. The two-dimensional (2D) and three-dimensional (3D) distribution of shale gas can only be predicted using seismic data. The quantitative relationship between porosity and P-wave velocity was obtained through the abovementioned petrophysical analysis (Figure 4.9), and the seismic prediction model or equation for the porosity of the study area was established as follows:

$$POR = 5 \times 10 v_p^{-2.745} \tag{4.8}$$

where $\nu_P$ is P-wave velocity.

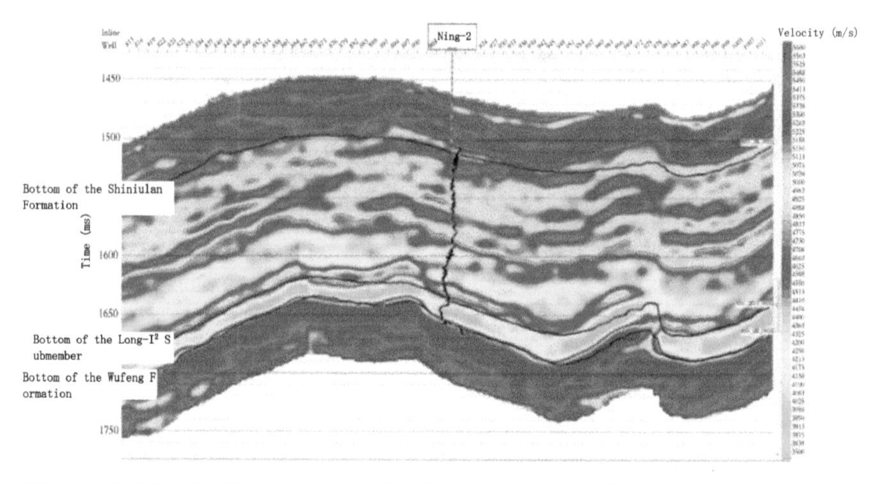


**Figure 4.10.** Section of pre-stack seismic inversion of the P-wave velocity at the location of well Ning-209.

According to Equation (4.8), the P-wave velocity data volume derived from the simultaneous pre-stack inversion can be converted into a porosity data volume to predict the 2D and 3D distributions of the porosity. A section of the pre-stack inversion of the P-wave velocity (Figure 4.10) shows that the P-wave velocity of the Wufeng–Longmaxi Formations decreases overall from top to bottom and that the interval from the bottom of the Wufeng Formation to the Long-$I_1$ Submember has the lowest P-wave velocity (green) of 3,400–3,700 m/s and is continuously distributed in the transverse distribution. The section of the predicted porosity (Figure 4.11) shows that the porosity of the Wufeng–Longmaxi Formations increases overall from top to bottom and that the bottom of the Wufeng Formation and the Long-$I_1$ Submember has the highest porosity (blue-purple) of 2–7% with some heterogeneity. In particular, the bottom of the Wufeng Formation and the bottom of the Long-$I_1$ Submember have the highest porosity of 5–7%, with a relatively continuous lateral distribution. The upper part of the Long-$I_1$ Submember has a porosity of 2–5%, which features relatively strong lateral heterogeneity. For well Ning-209, the results of the seismic prediction of porosity are consistent with the log interpretation results, indicating that the predicted porosity is reliable.

Based on the above results, the planar distribution map of the predicted average porosity of the Wufeng Formation and the Long-$I_1$

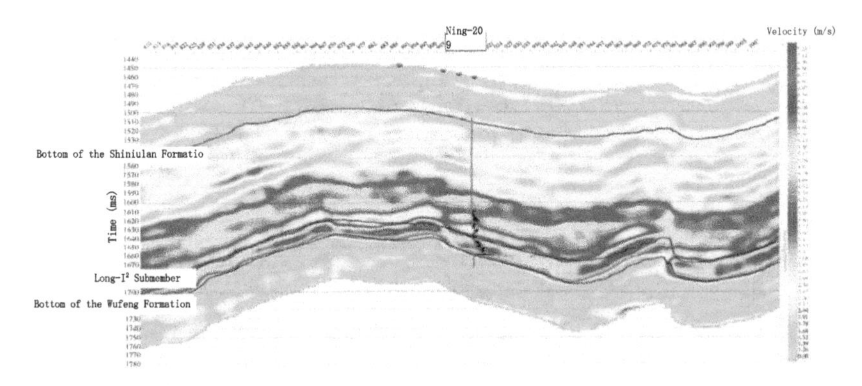


**Figure 4.11.** Seismic prediction results of porosity at the location of well Ning-209.

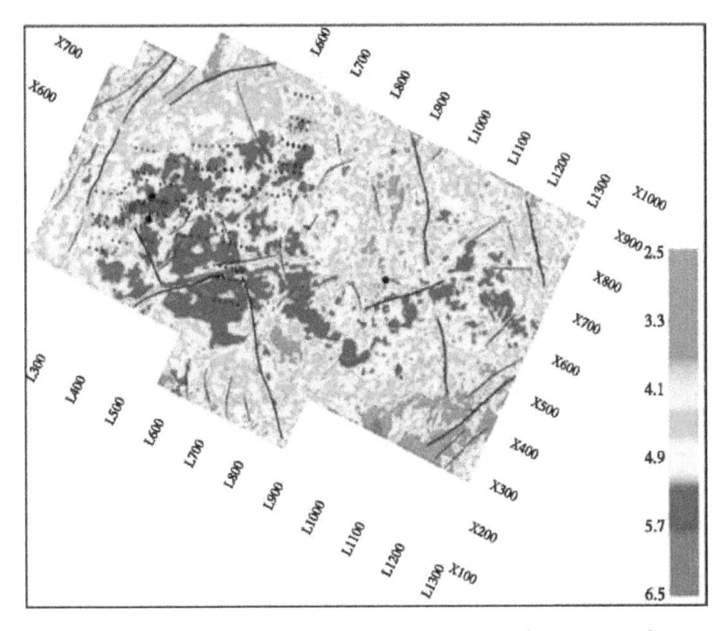


**Figure 4.12.** Map showing the planar distribution of porosity of reservoirs in the Changning area as predicted from seismic data.

Submember was obtained using the porosity data volume derived from the pre-stack inversion. The southwestern, middle, and northeastern portions of the area have a relatively high average porosity of more than 5%, indicating favorable overall physical properties of the shale gas reservoirs in the study area (Figure 4.12).

## 4.4. Quantitative prediction of TOC content

### 4.4.1. *Spatial distribution of TOC content*

The pre-stack inversion results show that the shale gas reservoirs in the high-TOC content intervals in the Changning block all have very low P-wave velocity, S-wave velocity, and density. Moreover, the higher the TOC content, the more obvious the low anomalies of the P-wave velocity, S-wave velocity, and density. As shown in Figure 4.13, the black lines on the sections at well locations are the log curves of P-wave velocity, S-wave velocity, and density. The inversion results are in good agreement with log curves, proving the reliability of the inversion results.

As shown in Figure 4.13, there is vertically one set of strata with significantly lower P-wave velocity, S-wave velocity, and density near the bottom of the Longmaxi Formation (as indicated by the arrow on the figure). It can be inferred from the above analyses that these strata are a set of high-quality shale gas reservoirs with high TOC content. Laterally, high-quality shale reservoirs are widely distributed and have good continuity, indicating a good prospect for shale gas development in this area.

After the porosity and the density data volumes are obtained, the density data volume can be converted into a TOC content data volume using the quantitative prediction template of the TOC content in order to achieve the quantitative prediction of the 2D and 3D distributions of the TOC content. Figure 4.14 shows the TOC content prediction section of the well Ning-209.

The TOC content prediction results show that the Longmaxi Formation has generally high TOC content of almost always above 1%, which tends to increase gradually toward the bottom. There is one set of strata with a TOC content higher than 3% at the bottom of the Longmaxi Formation, which is the main target of shale gas development in this area. The curve displayed at the well location is the TOC content curve based on log interpretation. The comparison shows that the TOC content prediction results are in good agreement with the log interpretation results, reflecting the reliability of the prediction results.

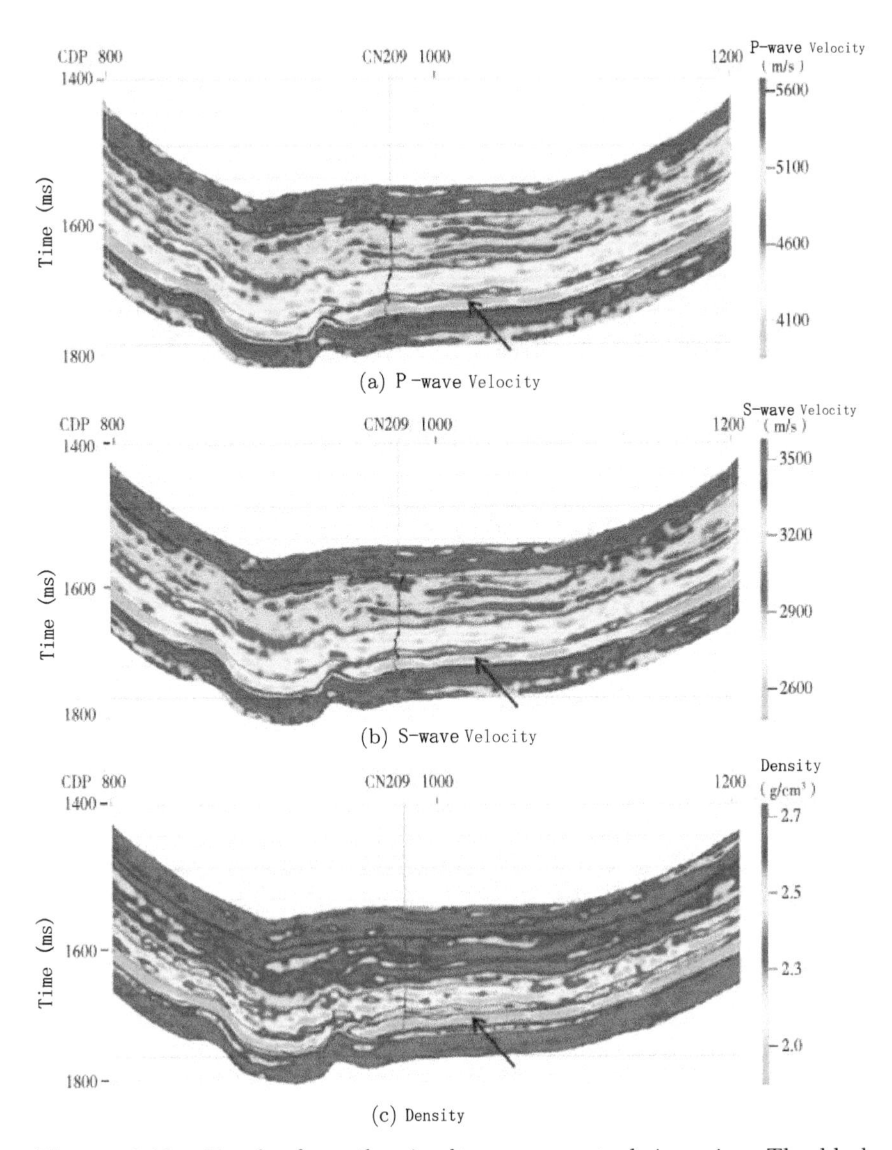


**Figure 4.13.** Results from the simultaneous pre-stack inversion. The black arrows indicate the target strata.

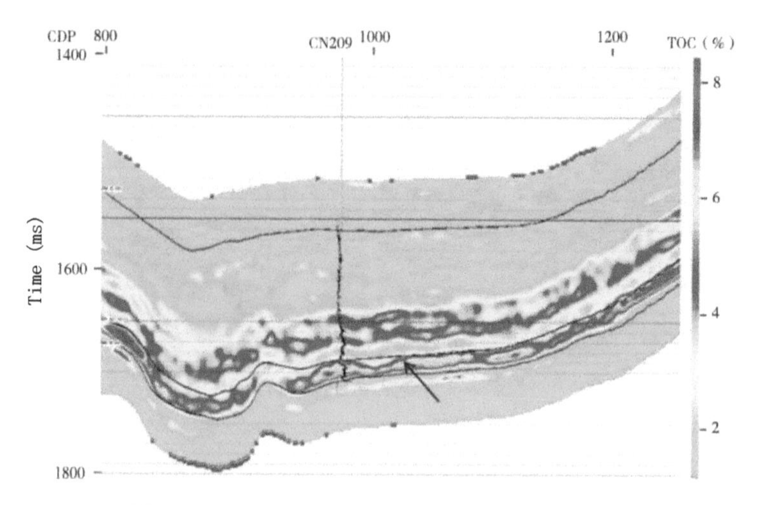


**Figure 4.14.** TOC content prediction results.

### 4.4.2. *Planar distribution characteristics of TOC content*

The TOC content planar prediction results show that the Changning area has relatively high TOC content and that the regions with high TOC content are continuously distributed, indicating a good development prospect in this area. Figure 4.15 shows the average TOC content distribution map of the Long-$I_1$ Submember in this block calculated using the top and bottom of Long-$I_1$ Submember as the time window. According to this figure, the Long-$I_1$ Submember has higher average TOC content of almost always above 2%, with a maximum of 6% and an average of 4%, suggesting a wider presence of sweet spots. With regards to lateral variations, the favorable zones tend to concentrate in the syncline core in the middle of the block. The syncline limbs around the block have relatively low TOC content, indicating that the TOC content distribution in this area is related to structures.

### 4.4.3. *Thickness distribution characteristics of high-quality reservoirs*

Adequate thickness and distribution area of shales with high TOC content are necessary to meet the requirements of industrial gas

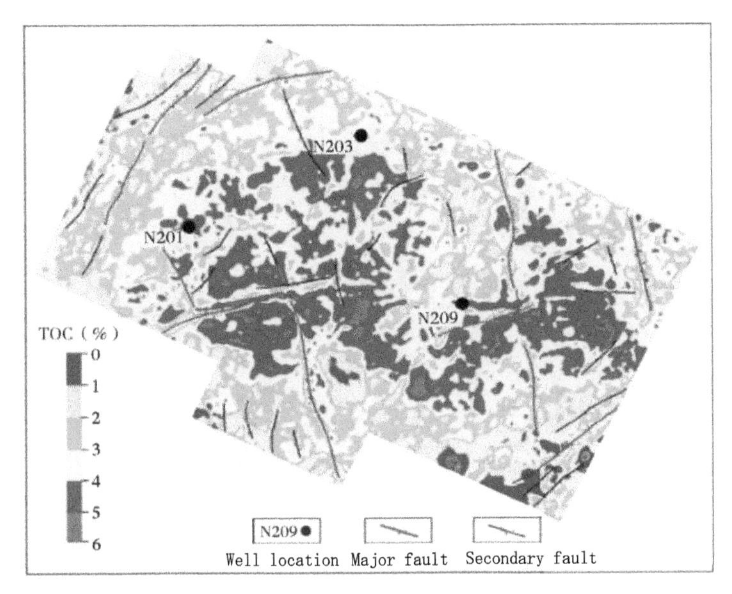


**Figure 4.15.** Map showing the planar distribution of predicted TOC content.

reservoir developments and are important conditions to ensure that shale gas reservoirs have sufficient organic matter and reservoir volume. Under equal conditions of organic matter content and gas generation intensity, thicker shales are associated with more gas. However, to form large-scale shale gas reservoirs, the thickness of shale must be greater than the effective thickness of hydrocarbon expulsion and should be greater than 30 m in general. Therefore, the thickness distribution of the strata with different TOC content values was predicted in this study.

Figures 4.16 and 4.17 show the predicted isopach maps of shale gas reservoirs with TOC content above 2% and 3%, respectively. These maps are derived from the predicted TOC content data volume. Figure 4.16 shows that high-quality reservoirs with TOC content above 2% have a thickness of 22–50 m, averaging 32.5 m, with thicker reservoirs mainly located in the middle and east-central portions of the study area.

Figure 4.17 shows that high-quality shale gas reservoirs with TOC content above 3% have a thickness of 2–26 m, averaging 14 m, with thicker high-quality reservoirs primarily distributed in the middle and east-central portions of the study area.

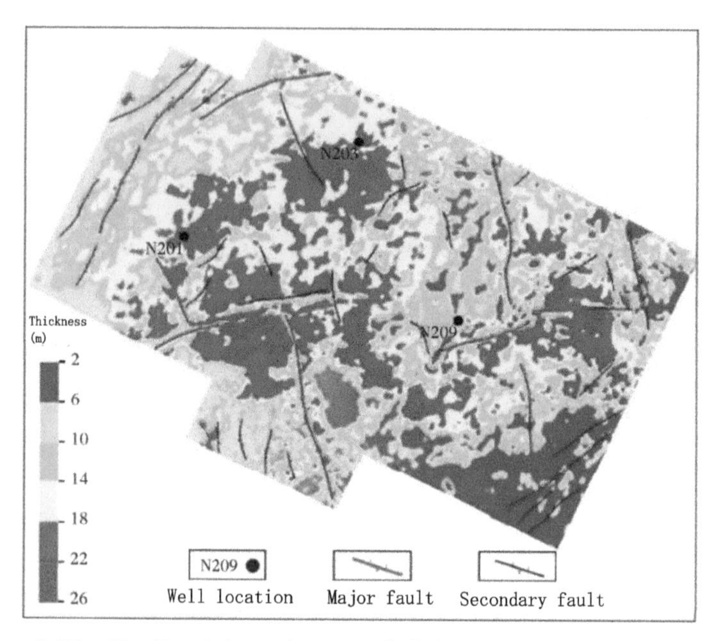


**Figure 4.16.** Predicted isopach map of shale reservoirs with TOC content greater than 2%.

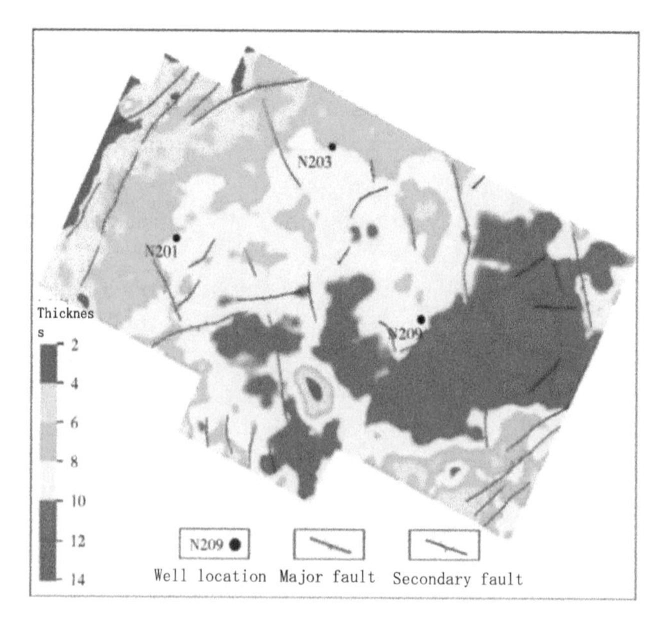


**Figure 4.17.** Predicted isopach map of shale reservoirs with TOC content greater than 3%.

The comparison between Figures 4.16 and 4.17 shows that the shale gas reservoirs of different quality show overall consistent distribution trends. However, there are obvious differences between them. The thicker reservoirs with TOC content above 2% (Figure 4.16) are mainly distributed in the east of the work area, with a thickness of 36–50 m, while the reservoirs in the middle part have a thickness of 29–36 m. The thicker reservoirs with TOC content above 3% (Figure 4.17) are mainly distributed in the middle portion of the study area, with a thickness of 18–26 m, while the reservoirs with a thickness of 18–26 m in the east are very limited. At present, the reason for these differences is unknown.

The TOC content distribution in the study area is controlled by the burial depths of structures and the fault distribution. Figure 4.18 is a map showing the burial depth and the fault distribution of the

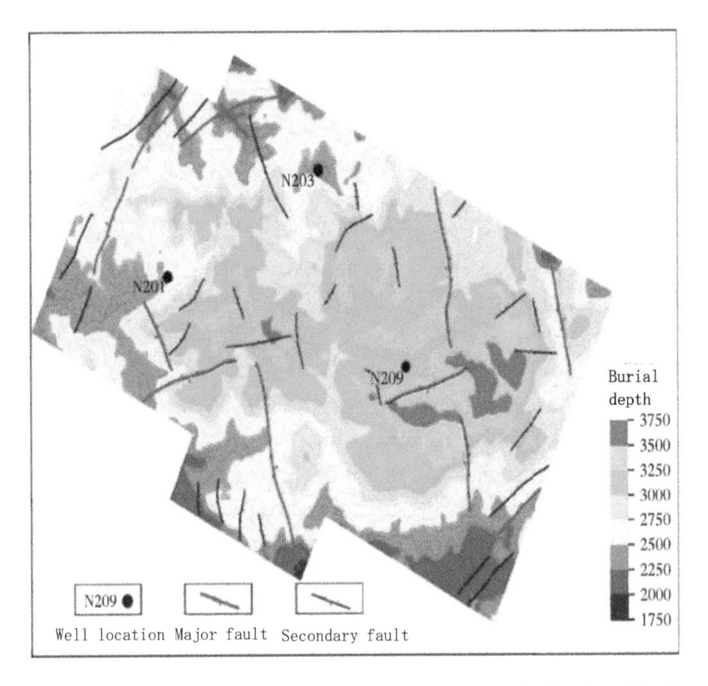


**Figure 4.18.** Map showing the burial depth and the fault distribution of the reservoirs in the Changning block.

*Note*: The colors represent the burial depth of reservoirs, with warm colors representing small burial depths and cool colors representing the large burial depths; the black lines are the contour lines for the burial depth.

target strata, in which the red lines indicate the first-order faults and the black lines indicate the second-order faults. It can be seen from the map that the work area is a syncline with an NE-oriented long axis, and the core of the anticline is in the central-east region of the block. By comparing Figures 4.16–4.18, the distribution pattern of high TOC content is roughly consistent with the syncline geometry, indicating that structures in this area play an important role in controlling the distribution of TOC content.

Faults also play an important role in controlling the distribution of TOC content. The distribution of faults in Figure 4.18 shows that the faults in this area are divided into two sets, one oriented roughly NS and the other roughly EW, dividing the area into several fault blocks. The TOC is low near the faults but high inside the fault blocks, indicating that they have a destructive effect on TOC content.

### 4.4.4. *Analysis of the prediction effects*

There are only two vertical wells in the study area, i.e., N201 and N203, and horizontal wells. Therefore, the data on well N209 were used as constraints in the inversion process. By contrast, data on well N203 and horizontal wells were used as control and did not participate in the inversion process. The drilling and test results show that the TOC content increases from the area at well N203 location in the northeast of the study area to the area at well N201 location in the middle and then to the southwest. The inversion results are consistent with the actual situation.

The map of the predicted average TOC content of the Long-$I_1$ Submember of the Longmaxi Formation is shown in Figure 4.15. According to this figure, the predicted average TOC content of the study area is higher than 2%, indicating the continuous distribution of high-quality shale reservoirs. The predicted average TOC content at well N201 location is 3.77%, and that at well N203 location is relatively low at 2.82%. The horizontal wells on wellpads H4 and H2 yielded the average TOC content of the horizontal section. As shown by the comparison of the prediction results with the production tests and log interpretation results (Table 4.2), the predicted average TOC content of high-quality shale reservoirs is roughly consistent with the log interpretation and production tests results, with a relative error of less than 3%, showing a high prediction accuracy. Moreover, higher

Table 4.2. Verification of the TOC content prediction results.

| Well | Logging interpretation TOC content (%) | Seismic prediction TOC content (%) | Relative error (%) | Test production ($10^4$ $m^3/d$) |
|---|---|---|---|---|
| N201 | 3.5 | 3.8 | 8.6 | 2.20 |
| N203 | 2.8 | 2.7 | 3.6 | 0.85 |
| H4-1 | 4.0 | 4.1 | 2.5 | 22.47 |
| H4-2 | 3.5 | 3.7 | 5.7 | 23.68 |
| H4-3 | 3.8 | 3.8 | 0.0 | 20.56 |
| H4-4 | 3.3 | 3.1 | 6.1 | 23.11 |
| H4-5 | 2.82 | 2.5 | 11.3 | 17.51 |
| 114-6 | 2.8 | 2.4 | 14.3 | 18.2 |
| H2-2 | 1.52 | 1.8 | 18.4 | 3.82 |
| 12-3 | 1.98 | 2.1 | 6.1 | 5.70 |
| 12-5 | 1.5 | 1.7 | 13.3 | 1.00 |
| 12-6 | 2.1 | 2.3 | 9.5 | 6.64 |

predicted average TOC content corresponds to higher production in the test production.

## 4.5. Seismic prediction of gas content in shale gas reservoirs

### 4.5.1. *Measurement method of gas content in shale gas reservoirs*

The prediction of sweet spots is a critical step in the exploration and development of shale gas reservoirs. The shale gas sweet spots refer to the areas with a moderate burial depth, high TOC content, high maturity, high gas content, and high content of brittle minerals, as well as well-developed natural fractures, where fracture networks are prone to form in the process of fracturing. They are the first choice for high and stable shale gas production and effective shale gas development. As an important component of sweet spots, gas content plays an important role in the commercial development of shale reservoirs.

The interpretation of log data can accurately characterize the vertical distribution characteristics of gas content at well locations, while the prediction of the planar distribution of gas content in reservoirs mainly relies on seismic data. However, seismic prediction of gas content of shale gas reservoirs is still immature at present.

Based on core analysis results and log data, Dodds *et al.* (2007) analyzed the seismic attributes of shale reservoirs, established the relationships between the seismic attributes and gas saturation using neural networks, and finally obtained gas saturation of shale reservoirs with a certain accuracy based on seismic attributes. Ross *et al.* (2009) analyzed pore structure, TOC content, and pore fluid composition of shale reservoirs in detail. They concluded that gas content was closely related to porosity and TOC content, and roughly determined a quantitative relationship between gas content, porosity, and TOC content based on log data. Bustin (2012) systematically studied the conditions and main controlling factors of the high production of shale gas by means of numerical simulation and indirectly obtained the main controlling factors of the gas content in shale reservoirs. Her findings show that the main geological factors controlling high shale gas production include fracture porosity, matrix porosity, Young's modulus, and Poisson's ratio and that the high shale gas production intervals have significantly high Young's modulus and low Poisson's ratio. This study indirectly indicates that gas content is closely related to two seismic physical parameters of rocks, namely, Young's modulus and Poisson's ratio. Altowairqi *et al.* (2015) conducted a systematic quantitative study of the influence of TOC content, gas content, and other parameters on the seismic elastic parameters under the condition of formation pressure, by conducting measurements of synthetic rock samples. The results show that as the TOC content and the gas content increase, the P-wave velocity, the S-wave velocity, and the density all decrease significantly. The quantitative relationships between gas content and TOC content and P-wave velocity, S-wave velocity, and density were also established.

In recent years, many Chinese researchers have conducted feasibility studies and demonstrations on the seismic prediction of gas content in shale reservoirs and carried out experiments using actual data volumes, achieving some results. Li *et al.* (2011) discussed the occurrence state of gas in shale reservoirs, analyzed the main factors influencing gas content, and preliminarily expounded on the role of gas content in the comprehensive evaluation of shale gas reservoirs. However, they did not study the prediction method of gas content.

Sun (2013) analyzed the correlation between seismic attributes and gas content and selected the three most sensitive attributes of gas content, i.e., average instantaneous frequency, arc length, and average energy. Using a multiple linear regression method, she established a prediction model of the gas content of shale reservoirs in the study area and accordingly predicted the gas content of Silurian high-quality shale reservoirs at the locations of the exploration wells. This method is convenient and fast, but its physical significance is yet to be clarified and suffers the multiplicity of solutions. Aiming at the shale gas reservoirs in the Jurassic Dongyuemiao Member in the Jiannan area, Liu *et al.* (2014) determined the lithologic associations and the seismic response characteristics of high-quality shale intervals by combining drilling and log data, predicted changes in the lithology and lithofacies of the study area using the waveform classification technique, predicted gas content of high-quality shales using the spectrum attenuation technique, and selected the most favorable gas-bearing area by combining the waveform classification prediction results. This method is highly comprehensive but lacks targeted analysis of the spectral response characteristics of shale gas reservoirs. Therefore, it can only be used for the qualitative prediction of gas content. Guo *et al.* (2015) analyzed the cross plots of gas content obtained from core analysis and geophysical elastic parameters and concluded that the most sensitive geophysical parameter of gas content is density. He established a mathematical model linking gas content and density and obtained a gas content data volume using this model based on the density volume derived from the pre-stack inversion. Moreover, he quantitatively predicted the gas content, achieving encouraging results. However, this method predicts the gas content only using a single parameter, i.e., density, which can be risky and carries high uncertainty.

Currently, studies worldwide on gas content mainly focus on the analysis and interpretation of log data of individual wells and core analysis. These methods can be used to accurately predict the vertical distribution characteristics of gas content at well locations. Their spatial prediction results have higher accuracy for areas with wells but have large errors for the areas without wells. Traditional gas content prediction techniques for reservoirs (e.g., AVO, spectrum decomposition, and analysis of seismic attributes) are mainly employed in China. However, these techniques remain in the stage of qualitative prediction, have low prediction accuracy, and lack univocal solutions, failing to meet the requirements of shale gas developments. At present, only the tectonic and reservoir thickness prediction results have been obtained in the study of the Changning block, without planar prediction results of gas content.

### 4.5.2. *CNN-based seismic prediction of gas content in shale gas reservoirs*

To quantitatively predict gas content in shale reservoirs, this study determined the spatial distribution of shale reservoirs with high gas content in the key area and then predicted the gas content using log combined with seismic data. All operations in this study were performed using the HampsonRussell software. The specific research approach is as follows:

(1) Quality control and preprocessing were conducted for the log and seismic data. For example, invalid data and null values were removed from log data, and the seismic data were denoised through static correction, velocity analysis, normal moveout correction, and pre-stack time migration (PSTM) to satisfy the inversion requirements. (2) The log and seismic data were interpreted. Key evaluation parameters for shale reservoirs such as the gas content at well locations were obtained through the interpretation of log data. The planar distributions of the top and bottom of the Long-$I_1$ Submember were determined through the interpretation of seismic data to clarify tectonic structures. (3) The seismic petrophysical analysis was carried out. On the one hand, the petrophysical characteristics of shale reservoirs and the vertical distribution of the shale reservoirs with high gas content were determined. On the other hand, the elastic parameters sensitive to gas content were selected. (4) The pre-stack seismic inversion was performed to obtain the seismic elastic parameters related to gas content. (5) A multivariate stepwise regression analysis was conducted to determine the type of elastic parameters and the optimal combination of parameters for predicting gas content. (6) The quantitative relationship between the gas content and the optimal combination of parameters was established to predict the gas content. (7) The results were analyzed to determine the spatial distribution of shales with high gas content. The workflow of the research approach is shown in Figure 4.19.

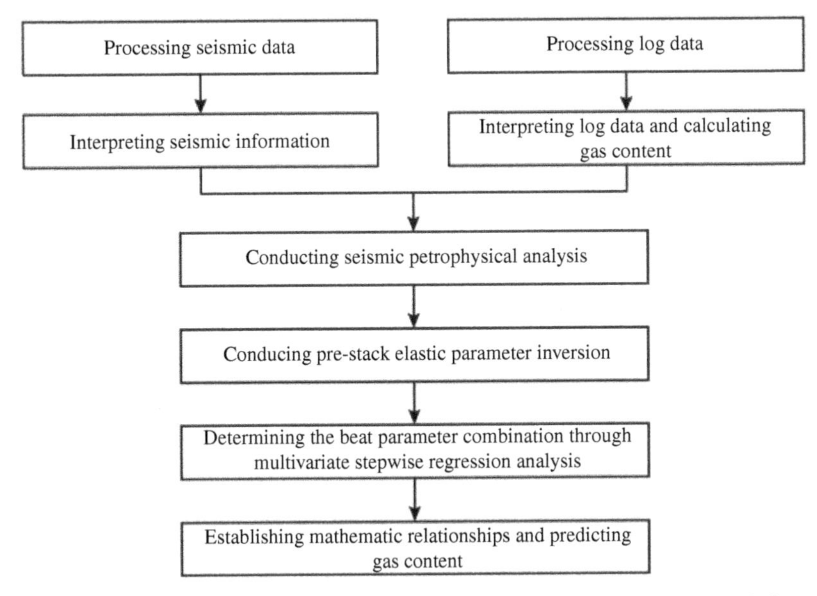


**Figure 4.19.** Workflow for seismic prediction of gas content in shale gas reservoirs.

Both the seismic inversion and the analysis of seismic attributes provided basic data for this approach. There are many types of seismic inversion, one of which is the pre-stack elastic wave impedance inversion (also referred to as elastic parameter inversion). The pre-stack elastic wave impedance inversion uses different theoretical basis, used data, methods, steps, and analysis criteria from the post-stack acoustic impedance inversion. It is based on the theories of seismic reflection and transmission. In other words, the amplitude of seismic reflections on an interface is not only related to the seismic elastic parameters of the media on both sides of the interface but also changes with the incident angle. Therefore, it can make full use of gather data of different offsets as well as log data such as S-wave, P-wave, and density to determine various elastic parameters related to the lithology and the hydrocarbon content through joint inversion and to comprehensively determine the physical properties and the hydrocarbon content of reservoirs.

Since the pre-stack elastic wave impedance inversion utilizes a large amount of seismic and log data, the multi-parameter analysis results are much more reliable than the results of the post-stack acoustic impedance inversion and can be used for the semi-quantitative and quantitative analysis of the hydrocarbon-bearing properties of hydrocarbon reservoirs. Seismic attributes refer to the geometric, kinematic, dynamic, or statistical characteristics of seismic waves, which are derived from the mathematical transformation of pre- or post-stack seismic data. There are many kinds of seismic attributes, which reflect the structure, strata, physical properties, and hydrocarbon content of hydrocarbon reservoirs from different perspectives. Therefore, they are widely used in the characterization and prediction of hydrocarbon reservoirs. Many researchers have mentioned the significance of seismic attributes for the study of reservoir characteristics. Hampson *et al.* (2001) defined the seismic attributes as any mathematical transformation results of seismic trace data, mainly including amplitude, waveform, frequency, energy, phase, and correlation analysis. In practice, seismic attributes can be extracted from seismic traces near wells on original seismic profiles, and they are internal seismic attributes. Moreover, seismic attributes can also be extracted from other processed data volumes, such as wave impedance inversion attribute, and coherence attribute. They are external seismic attributes. Before seismic attribute extraction, it is necessary to establish the corresponding time-to-depth conversion relationship between the target parameters and the seismic attributes, and the time window should be focused on the target intervals, that is, the location of the target intervals should be finely calibrated through a synthetic seismogram. In the prediction of gas content, relevant seismic attributes are input as external data for the selection of the optimal sensitive parameters, training of neural networks, and prediction of gas content.

To predict the gas content of shale reservoirs, this study input the following data into the Emerge module in the HampsonRussell software: (1) external attributes, which are elastic parameters such as density, Young's modulus, and Poisson's ratio derived from the pre-stack inversion; (2) internal attributes, which include raw seismic data and the seismic attributes including the instantaneous frequency extracted from the raw data, and (3) the target parameter, i.e., the gas content calculated from log data (Figure 4.20). The Emerge module can be used to establish the relationships between gas content and seismic attributes and to calculate a gas content data volume in three steps: (1) the single-attribute analysis is conducted to determine the

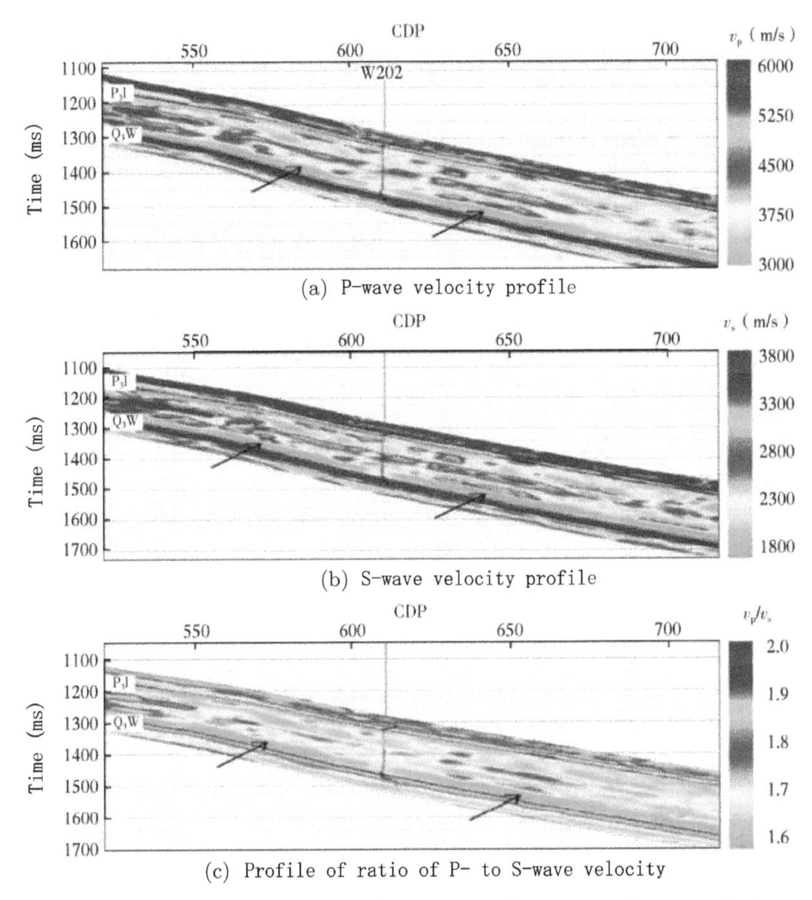


(a) P-wave velocity profile

(b) S-wave velocity profile

(c) Profile of ratio of P- to S-wave velocity

**Figure 4.20.** Inversion results of the P-wave velocity, the S-wave velocity, and the ratio of P- to S-wave velocity. The inserted curves refer to log data, and the arrows indicate the target intervals.

seismic attributes closely related to the target parameter (gas content); (2) the relationships between the seismic attributes and the target parameter are established using linear or nonlinear algorithms, and (3) the relationships are applied to the whole 3D block to obtain the data volume of the target parameter (gas content).

The single-attribute analysis based on machine learning aims to extract and screen seismic attributes and then calculate the correlations between the target parameter and individual seismic attributes. Attribute extraction is mainly designed to extract common attributes including integral of amplitude envelope, apparent polarity, trace integral, and instantaneous phase from raw seismic data volumes. Then, a series of mathematical transformations such as reciprocal and logarithm are performed for these internal and external attributes. Finally, the correlations between the transformed seismic attributes and the target parameter are calculated and sorted. Table 4.3 shows the results of the single-attribute regression analysis for various seismic attributes. The ranking in this table is determined based on the correlation and the prediction errors. A higher single-attribute prediction error corresponds to a weaker correlation. The analysis results show that density is closely correlated with gas content and yields a small prediction error. Therefore, density is regarded as the most important external attribute for predicting gas content.

The combination of multiple attributes can significantly improve the prediction accuracy of the target attribute. The multi-attribute stepwise regression based on the single-attribute regression analysis is a simple and practical prediction technique, which can be used to predict the target attribute according to the input attributes. Its ultimate purpose is to determine the combination of seismic attributes for predicting gas content and establish the relationship between the gas content and the optimal combination of attributes.

Table 4.4 shows the prediction errors of different combinations of attributes. The training error decreases with an increase in the number of input attributes. However, this does not mean that the more the attributes involved, the more realistic the prediction results. Therefore, this study stopped adding attributes after the optimal number of attributes was determined through test-based verification. The test for verification was carried out in stages, and the training wells and attribute 1 were used as the optimal individual attributes to calculate the gas content log curves of verification wells. This test stopped when all the wells were used for training. At each stage, the error between the predicted value and the actual value was calculated as the verification error. Afterward, the above stage was repeated to define the optimal two, three, or more attributes. Then, the optimal number of attributes was determined according to the relationship between the verification error and the number of attributes. As shown in Figure 4.21, as the number of attributes increased, the prediction error decreased and the verification error first decreased and then increased and finally tended to stabilize.

**Table 4.3.** Error analysis of seismic multi-attribute prediction of gas content.

| Target parameter | Attribute | Error | Correlation coefficient | Ranking |
|---|---|---|---|---|
| Gas content | Density (inverted) | 0.014113 | −0.568878 | 1 |
| Gas content | Square of Poisson's ratio | 0.014770 | −0.509040 | 2 |
| Gas content | Logarithm of P-wave velocity | 0.014770 | −0.509040 | 3 |
| Gas content | P-wave impedance | 0.014923 | −0.493665 | 4 |
| Gas content | Average frequency | 0.015969 | 0.366020 | 5 |
| Gas content | Trace integral | 0.016315 | −0.309900 | 6 |
| Gas content | Product of Lambda constant and density | 0.016046 | −0.293164 | 7 |
| Gas content | Absolute values of amplitude | 0.016414 | 0.291613 | 8 |
| Gas content | Bulk modulus | 0.016457 | −0.283254 | 9 |
| Gas content | Curvature | 0.016628 | 0.247046 | 10 |
| Gas content | Reciprocal of instantaneous frequency | 0.016844 | −0.243381 | 11 |
| Gas content | Reciprocal of instantaneous amplitude | 0.016728 | 0.222990 | 12 |
| Gas content | Integral | 0.016763 | −0.213914 | 13 |
| Gas content | Apparent polarity | 0.016801 | 0.203571 | 14 |
| Gas content | Amplitude weighted phase | 0.016840 | −0.192123 | 15 |
| Gas content | Instantaneous frequency | 0.016861 | 0.185843 | 16 |
| Gas content | Filtering 5/20–25/30 | 0.016864 | 0.184831 | 17 |
| Gas content | Dominant frequency | 0.016877 | −0.180857 | 18 |
| Gas content | Amplitude weighted phase cosine | 0.016897 | 0.174481 | 19 |
| Gas content | Original seismic | 0.016898 | 0.174136 | 20 |
| Gas content | High-pass filtering | 0.016898 | 0.174136 | 21 |
| Gas content | Amplitude weighted frequency | 0.016903 | 0.172463 | 22 |
| Gas content | Amplitude envelope | 0.016907 | 0.171082 | 23 |
| Gas content | Instantaneous phase | 0.016924 | −0.165213 | 24 |
| Gas content | Filtering 5/10–15/20 | 0.016996 | 0.137829 | 25 |
| Gas content | Cosine of instantaneous phase | 0.017035 | 0.120588 | 26 |
| Gas content | S-wave impedance | 0.017072 | −0.101385 | 27 |
| Gas content | Product of shear modulus and density | 0.017088 | −0.091475 | 28 |
| Gas content | S-wave velocity | 0.017095 | −0.087058 | 39 |
| Gas content | Filtering 5/60–65/70 | 0.017106 | −0.079529 | 30 |
| Gas content | Filtering 5/30–35/40 | 0.017147 | 0.038762 | 31 |
| Gas content | Second-order reciprocal | 0.017158 | −0.014951 | 32 |
| Gas content | Filtering 45/50–55/60 | 0.017158 | −0.013665 | 33 |
| Gas content | Filtering 35/40–45/50 | 0.017159 | −0.009266 | 34 |

**Table 4.4.** Error analysis of seismic multi-attribute prediction of the gas content.

| Target parameter | Optimal attribute | Prediction error | Verification error |
|---|---|---|---|
| Gas content | Density (inverted) | 0.015358 | 0.0326 |
| Gas content | Square of Poisson's ratio | 0.013949 | 0.032 |
| Gas content | Logarithm of P-wave velocity | 0.012184 | 0.03 |
| Gas content | P-wave impedance | 0.011562 | 0.028 |
| Gas content | Average frequency | 0.010045 | 0.021 |
| Gas content | Trace integral | 0.010039 | 0.0208 |
| Gas content | Product of Lambda constant and density | 0.009178 | 0.021 |

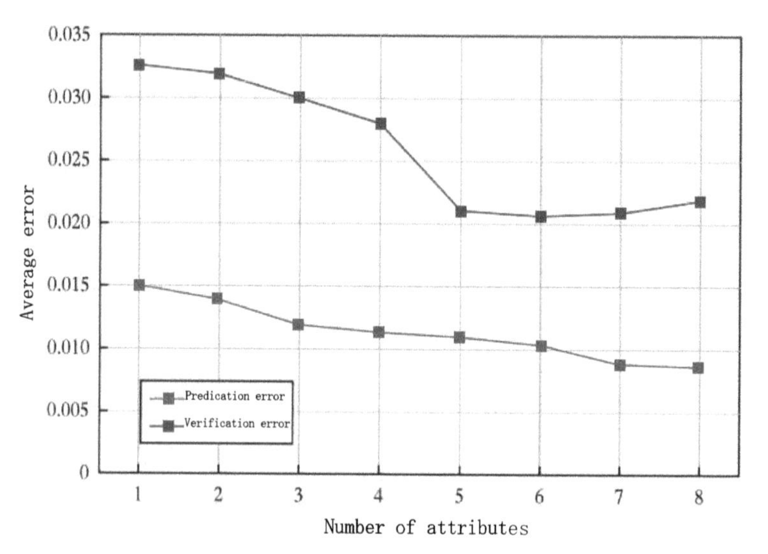


**Figure 4.21.** Error analysis of seismic multi-attribute prediction of gas content.

The optimal number of attributes occurred when the verification error started to increase or stayed unchanged. When the seventh attribute was added, the verification error increased. The verification error of six seismic attributes was the smallest at 0.0208–0.021, indicating that the first six attributes were the optimal inputs for the prediction of gas content in shale gas reservoirs in the study area. The optimal combination of attributes and the error analysis results are shown in Table 4.4. It can be seen from Table 4.4 and Figure 4.21 that the optimal combination of attributes for predicting the gas content consisted of density (Inv-DEN), the square of Poisson's ratio [$(\text{Inv-PR})^2$], the logarithm of the P-wave velocity [lg(Invip)], P-wave impedance (Inv-AI), average frequency, trace integral, and Lambda-Rho (product of Lambda constant and density). Afterward, the relationships between gas content and these attributes were established.

Finally, the 3D gas content seismic data volume can be calculated. Specifically, based on the optimal number and types of attributes determined above, the final multivariate regression relationship was applied to the whole 3D data volume to calculate the 3D gas content seismic data volume. Figure 4.22(b) shows a section extracted from the gas content data volume of the Longmaxi Formation obtained in this study, with the gas content calculated using the log data as the input data. A strong correlation can be observed at well locations.

The results of this study show that the density of the shale gas reservoirs is closely related to the gas content and plays an important role in its prediction. Figures 4.22(a) and 4.22(b) show the sections of inverted density and predicted gas content at the location of well W202, revealing that there is a strong negative correlation between the density and the gas content. In other words, higher gas content is associated with significantly lower reservoir density.

Based on the shale gas production and surrounding prospects of the Changning block, this study proposed an evaluation scheme of the shale reservoirs in the block using the fuzzy comprehensive evaluation method and referenced previous research results of this block, as shown in Table 4.5. According to the scheme and the planar distribution of gas content in the Changning block obtained with seismic prediction, reservoirs in the block can be divided into three types, namely, Class-I, -II, and -III. The Class-I reservoirs have gas content greater than 3 $\text{m}^3/\text{t}$ and density less than 2.2 $\text{g/cm}^3$, the Class-II reservoirs have gas content of 1.5–3 $\text{m}^3/\text{t}$ and density greater than 2.2 $\text{g/cm}^3$, and the Class-III reservoirs have gas content less than 1.5 $\text{m}^3/\text{t}$ and density greater than 2.45 $\text{g/cm}^3$. The Class-I reservoirs have higher gas content and are the most favorable zones for shale gas development. The Class-II reservoirs have moderate gas content and are favorable reservoirs. The Class-III reservoirs have low gas content and are inconducive to shale gas development.

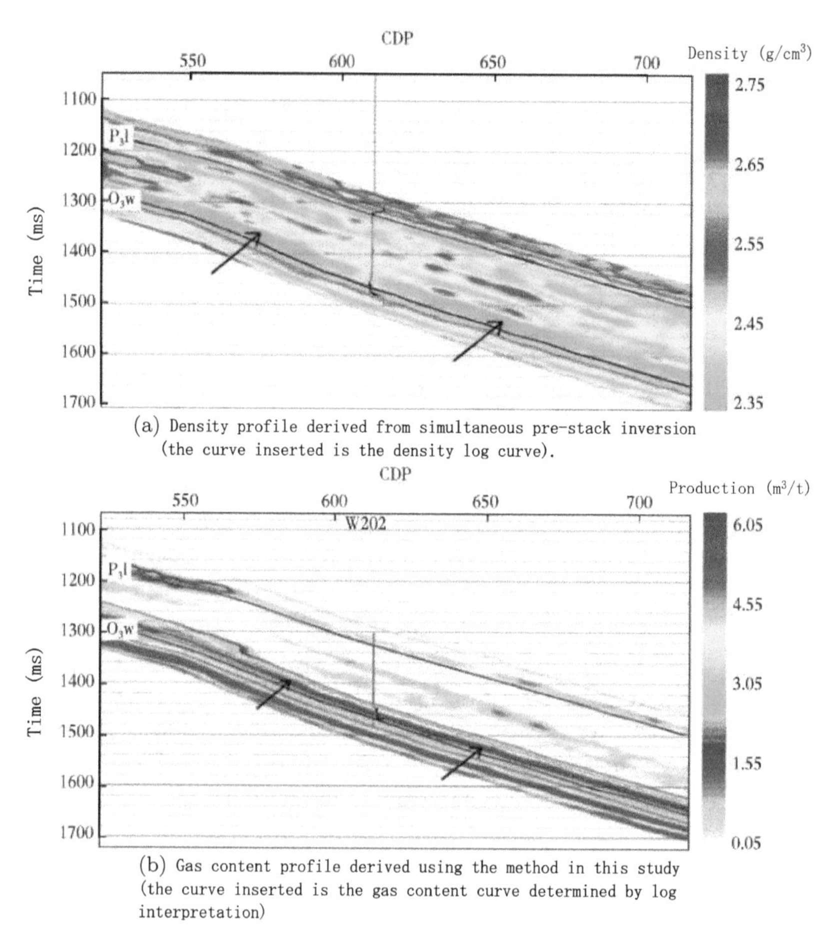


(a) Density profile derived from simultaneous pre-stack inversion (the curve inserted is the density log curve).

(b) Gas content profile derived using the method in this study (the curve inserted is the gas content curve determined by log interpretation)

**Figure 4.22.** Sections of inverted density and predicted gas content of shale gas reservoirs in the Longmaxi Formation.

**Table 4.5.** Classification scheme of sweet spots based on evaluation parameters including gas content.

| Reservoir category | Thickness (m) | TOC (%) | Porosity (%) | Gas content ($m^3/t$) | Density ($g/cm^3$) |
|---|---|---|---|---|---|
| Class-I | ≥30 | ≥3 | ≥5 | ≥3 | ≥2 |
| Class-II | ≥30 | 2–3 | 3–5 | 2–3 | 2–1.5 |
| Class-III | <30 | <2 | <3 | <2 | <1.5 |

**Table 4.6.** Production prediction and verification.

| Wellpad | Well name | Predicted value ($m^3/t$) | Test value ($m^3/d$) |
|---|---|---|---|
| H4 | 14–1 | 3.60 | 9.99 |
| | 14–2 | 4.15 | 21.60 |
| | 14–3 | 3.85 | 16.49 |
| | 14–4 | 3.75 | 14.50 |
| | 14–5 | 3.60 | 9.14 |
| | 14–6 | 3.70 | 14.31 |
| W201 | W201 | 2.60 | 1.10 |
| | W201-11 | 3.15 | 3.31 |

To verify the reliability of the results in this study, the latest production data from the latest horizontal wells in the Changning block were analyzed. As shown in Table 4.6 and Figure 4.23, eight wells have been drilled recently, belonging to wellpads H4 and W201. Wellpad H4 is located in the Class-I sweet spot area as shown in Figure 4.23. This wellpad has six horizontal wells, which have a production rate of $9.14 \times 10^4 - 21.6 \times 10^4$ $m^3/d$, with an average of $14.3 \times 10^4$ $m^3/d$. The predicted gas content of these six horizontal wells is 3.6–4.15 $m^3/t$, with an average of 3.78 $m^3/t$. The W201 wellpad has one vertical well and one horizontal well, both of which lie in the Class-II sweet spot areas as shown in Figure 4.23. The vertical well and the horizontal well (W201-H1) of wellpad W201 have production of $1.1 \times 10^4$ $m^3/d$ and $3.31 \times 10^4$ $m^3/d$, respectively, and predicted gas content of 2.6 $m^3/t$ and 3.15 $m^3/t$, respectively. Therefore, the field production data verified the reliability of this study.

In summary, the way the 3D seismic data are used to predict the spatial distribution and variation of gas content in shale gas reservoirs is of great significance for the exploration and development of shale gas in this area. This study presented a method for predicting the gas content using both seismic log data and evaluated two types of sweet spots. The Class-I sweet spots with high gas content are the best areas for production, and the Class-II sweet spots with medium gas content are areas for future shale gas production. The reservoirs in the key area of wellpad W201 can be divided into three categories. These results can guide the well location selection and industrial production of the study area. The research methods in

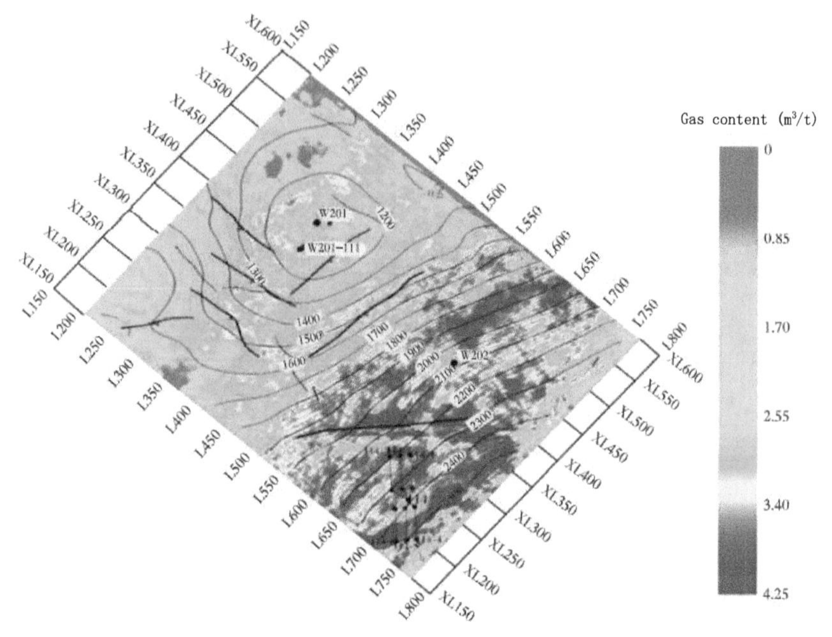


**Figure 4.23.** Planar distribution map of the average gas content in shale gas reservoirs in the Longmaxi Formation.

this chapter can be applied to other regions and provide a reference for the development of shale gas in China.

## 4.6. Summary

(1) In this chapter, the workflow for quantitative seismic prediction of TOC content was established through systematic research. Specifically, the seismic petrophysical analysis showed that density is the most sensitive to TOC content and that there is a high correlation between them. The quantitative relationship between TOC content and density was established by seismic petrophysical modeling to form the quantitative prediction template of TOC content. Based on this, a high-precision density data volume was obtained using the simultaneous pre-stack inversion, and then the quantitative prediction of TOC content was performed.

(2) The high-quality reservoirs with TOC content greater than 3% in the Changning block have a thickness of 2–26 m, with an average of 14 m. Thick high-quality reservoirs in this block are mainly distributed in the middle and central-east portions of the study area. Reservoirs with TOC content greater than 3% have roughly the same distribution trend as those with TOC content greater than 2%, but these reservoirs with different TOC contents have significantly different distribution patterns in local areas.
(3) The burial depth of structures and fault distribution play a significant role in controlling the planar distribution of the TOC content in the study area. The reservoirs with high TOC content are mainly distributed in the syncline core, with a distribution trend roughly consistent with the syncline morphology. The faults divide the Changning block into several fault blocks. The TOC content is low near the faults and is high inside the fault blocks. The faults destruct the high-quality reservoirs with high TOC content.
(4) Compared with the log interpretation and production test results, the predicted TOC content has a small error and high accuracy. The areas with high predicted TOC content have high production in the production test. This finding indicates that the prediction method is reliable for the exploration of shale gas in the Sichuan Basin. Therefore, this study provides an important means for the quantitative prediction of TOC content in shale gas reservoirs and is greatly significant for the efficient development of shale gas.
(5) As indicated by the exploration and exploitation practice of marine shale gas in southern China and practical data used in this study, there is a significant positive correlation between pore pressure and gas production of shale gas reservoirs, and the overpressure of strata is essential for high shale gas production in the Changning block. Therefore, the accurate prediction of formation pressure is very important for the scale-efficiency development of marine shales widely distributed in southern China.
(6) The prediction of 2D and 3D distributions of formation pressure mainly relies on seismic and log data. All the existing methods for the prediction of formation pressure have shortcomings. Methods developed based on normal compaction trend lines have limited applicability since accurate normal compaction trend lines

are difficult to obtain. Similarly, the pore pressure prediction method based on effective stress ignores the fact that P-waves are significantly affected by fluids, resulting in inaccurate pressure prediction results. Therefore, the current seismic prediction accuracy of formation pressure fails to meet the requirements of efficient shale gas development.

(7) This study deduces a new calculation formula for formation pressure based on geomechanics. Using the new calculation formula as well as the simultaneous pre-stack inversion technique, this study developed a new prediction method for formation pressure. The application results show that this prediction method has high accuracy and wide applicability and thereby is significant for the sweet spot prediction and well deployment of marine shales widely distributed in southern China.

## Bibliography

Altowairqi Y, Rezaee R, Evans B, *et al.* Shale elastic property relationships as a function of total organic carbon content using synthetic samples. *Journal of Petroleum Science and Engineering*, 2015, 133: 392–400.

Biot MA. Mechanics of deformations and acoustic propagation in porous media. *Journal of Applied Physics*, 1962, 33 (4): 1482–1498.

Biot MA. Theory of propagation of elastic waves in a fluid-saturated porous solid: Lowfrequency range. *Acoustical Society of America*, 1956, 28 (2): 168–178.

Castagna JP, Batzle ML, Eastwood RL. Relationships between compressional-wave and shear-wave velocities in clastic silicate rocks. *GeohPysics*, 1985, 50 (4): 571–581.

Chen Y, Jiang LC, Hu J, *et al.* A new method for quantitative prediction of fractured pores in shale reservoirs. *Geological Science and Technology Information*, 2018, 37 (1): 115–121.

Gassmann F. Elastic waves through a packing of spheres. *Geophysics*, 1951, 16: 673–685.

Guo X, Yin Z, Li J. Seismic quantitative prediction technology of gas content in marine shale and its application-taking Jiaoshiba area of Sichuan Basin as an example. *Petroleum Geophysical Exploration*, 2015, 50 (1): 144–149.

Hampson D, Schuelke J, Quirein J. Use of multi-attribute transform to predict log properties from seismic data. *Geophysics*, 2001, 66: 220–236.

Han DA, Nur A, Morgan D. Effect of porosity and clay content onwave velocities in sandstones. *Geophysics*. 1986, 51 (11): 2093–2107.

He Xilei. *Petrophysical Basis and Seismic Porosity Inversion for Hydrocarbon Prediction*. Chengdu: Chengdu University of Technology, 2012.

Li Y, Qiao D, Jiang W, *et al.* Overview of Shale gas gas content and geological evaluation of Shale gas. *Geological Bulletin*, 2011, 30 (2/3): 308–317.

Liu C, Yi J, Liu H. Study on earthquake prediction method of Shale gas in Jurassic Dongyue Temple section in Jiannan area. *Journal of Petroleum and Natural Gas*, 2014, 36 (10): 69–73.

Ma X, Hu Y, Wang F. Innovation and effect of integrated development of natural gas industry in Sichuan Basin. *Natural Gas Industry*, 2019, 39 (7): 1–8.

Mavko G, Mukerji T, Dvorkin J. *The Rock Physics Handbook: Tool for Seismic Analysis in Porous Media*. Cambridge University Press, Cambridge, UK, 2003.

Mavko G, Mukerji T. Seismic pore space compressibility and Gassmann's relation. *Geophysics*, 1995, 60 (6): 1743–1749.

Quirein JA, Schuelke JS, Hampson DP, *et al.* Use of multiattribute transform to predict log properties from seismic data. *Geophysics*, 2001, 66 (1): 220–236.

Raymer LL, Hunt ER, Gardner JS. An improved sonic transit time-to-porosity transform. In *Proceedings of the SPWLA 21st Annual Meeting*, Lafayette, Louisiana, July 1980, pp. 1–13.

Russell BH, Hedlin K, Hilterman FJ, *et al.* Fluid-property discrimination with AVO: A Biot-Gassmann perspective. *Geophysics*, 2003, 68 (1): 29.

Shi Y, Yao FC, Sun HS, *et al.* Seismic density inversion and formation porosity estimation. *Journal of Geophysics*. 2010, 53 (1): 197–204.

Su J, Qu D, Chen C, *et al.* Comparative analysis of prestack seismic inversion methods — Jiaoshiba Shale gas reservoir exploration example. *Petroleum Geophysical Exploration*, 2016, 51 (3): 581–588+418–419.

Sun Xiaoqin. Application of Seismic attribute analysis technology in prediction of shale gas content in Pengshui Block. *Petroleum Geology and Engineering*, 2013, 27 (4): 39–41.

Sun YF. Core-log-seismic integration in hemipelagic marine sediments on the eastern flank of the Juan De Fuca Ridge. *ODP Scientific Results*, 2000, 168: 21–35.

Sun YF. Pore structure effect on elastic wave propagation in rocks: AVO modeling. *Journal of Geophysics and Engineering*, 2004, 1 (4): 268–276.

Wang T, Zhao BY, Dai XF, *et al.* Prestack multi-attribute probabilistic neural network inversion of reservoir porosity. *Geophysical and Geochemical Exploration Calculation Technology*, 2013, 35 (02): 162–167+118.

Wyllie M, Gregory A, Gardner G. Elastic wave velocities in heterogeneous and porous media. *Geophysics*, 1956, 21: 41–70.

Yun MH, Yi WQ, Zhuang HY. Empirical relationship between elastic modulus of sandstone and porosity, shale content, effective pressure and temperature. *Petroleum Geophysical Exploration*, 2001, 36 (3): 308–314.

Zeng Q, Chen S, He P, *et al.* Seismic quantitative prediction of Shale gas sweet spot area of Longmaxi Formation in Weiyuan, Sichuan Basin. *Petroleum Exploration and Development*, 2018, 45 (3): 406–414.

Zhang JJ, Li HB, Yao FC. Rock critical porosity inversion and S-wave velocity prediction. *Applied Geophysics*, 2012, 9 (1): 57–64.

Zhang Yingbo. Biot Theory in application to seismic prospecting. *Geophysical Prospecting for Petroleum*, 1994 (4): 29–38.

Zhang Yong, He GS, Li YJ, *et al.* Neural network porosity prediction technology based on pre-stack inversion: A case study of the Nanchuan area. *Science and Technology and Engineering*, 2019, 19(25): 83–89.

Zhou Dan, Zhu T, Hu HF, *et al.* A porosity prediction method based on pre-stack inversion. *Geophysical and Geochemical Exploration Calculation Techniques*, 2015, 37 (4): 472–477.

Zhu Guangsheng. Use of P-wave and S-wave in porosity prediction with seismic data. *Journal of Jianghan Petroleum Institute*, 1992, 14(1): 27–32+39–38.

https://doi.org/10.1142/9789811283185_0005

# Chapter 5

# Seismic Prediction of Key Engineering Parameters of Shale Gas Sweet Spots

Key engineering parameters of shale gas sweet spots (e.g., reservoir brittleness) are key factors for the industrial production of shale gas. Moreover, they can be used as guidance and references in the design and optimization of fracturing construction schemes and assist in dealing with challenges concerning the schemes. This chapter mainly introduces the seismic prediction methods of reservoir brittleness, pore pressure, and *in-situ* stress as well as the principle and applications of the seismic prediction of natural fractures. In addition, this chapter makes comparative analyses and improvements in these methods.

## 5.1. Seismic prediction of reservoir brittleness

Industrial production of shale gas is possible only after hydraulic fracturing. Therefore, the brittleness of shale gas reservoirs is paramount for the size and the complexity of the fracture network created by hydraulic fracturing. Higher brittleness makes it more prone to artificially create large-scale complex fractures. Therefore, the identification of geological sweet spots for shale gas production, besides gas content, should also be driven by shale brittleness.

### 5.1.1. *Working principles*

Sweet spots are areas with high brittle mineral content where complex fracture networks are prone to form by hydraulic fracturing. As shown by the successful experience in shale gas exploration and exploitation at home and abroad, higher brittle mineral content in shale gas reservoirs makes it easier to form artificial fractures by hydraulic fracturing and to effectively stimulate the reservoirs. The brittleness of shale gas reservoirs can be frequently obtained from seismic prediction and the commonly used method is as follows. First, seismic inversion is employed to determine the rock mechanical characteristic parameters (e.g., elastic modulus and Poisson's ratio), which are then directly or indirectly used to evaluate the ability of shale gas reservoirs to form artificial fracture networks through hydraulic fracturing.

Many scholars at home and abroad researched the characterization of rock brittleness. At present, there are four methods to characterize the brittleness of shale gas reservoirs, as follows. (1) Direct measurement in the laboratory: the brittle minerals such as quartz in reservoirs are determined through mineral composition analysis, and then the reservoir brittleness is represented using the percentage of brittle minerals. (2) Brittleness index method: elastic parameters related to rock brittleness are obtained using rock mechanical formulas and are then used to directly or indirectly characterize reservoir brittleness. Among these elastic parameters, Young's modulus and Poisson's ratio are the most commonly used to calculate the brittleness index of rocks, which is the most commonly used to characterize rock brittleness; (3) Rock mechanical experiments: mechanical experiments (e.g., tension or compression) on reservoir rocks are conducted in the laboratory to obtain the stress–strain curves of the rocks, and then the reservoir compressibility is evaluated according to the characteristics of the stress–strain curves. (4) Large physical simulation or fracturing experiments: the fracturing process is simulated using large physical models in the laboratory or conventional fracturing experiments are performed on site to obtain data on rock compressibility. Many foreign researchers have carried out a lot of rock mechanical and fracturing experiments on the Barnett and the Woodford shales and have gained considerable insight. Generally, shale gas reservoirs with brittle mineral content greater than 40%

and brittleness index greater than 0.4 have favorable compressibility and will respond well to hydraulic fracturing.

Although rock composition analysis and rock mechanical experiments in the laboratory can be used to obtain accurate key rock mechanical parameters and accurately characterize the rock brittleness, their feasibility and time cost are rather unpractical. It is necessary to predict the rock brittleness to design an adequate fracturing procedure in shale gas development. However, compared to conventional natural gas reservoirs, it is more difficult and costly to collect cores of shale gas reservoirs. Therefore, despite the reliability of lab-based brittleness evaluation, its practical feasibility is restricted by many conditions. Given this, a set of methods for evaluating the brittleness of shale gas reservoirs have been established using geophysical logging and seismic data based on the shale gas exploration and development in North America. Table 5.1 shows various formulas for describing the brittleness of hydrocarbon reservoirs, among which $B_8$ and $B_9$ are most frequently used. $B_9$ is a brittleness evaluation method based on mineral composition. It is frequently used to describe the brittleness of shale gas reservoirs in log data interpretation on the premise that the mineral composition of a strata has been determined in advance. $B_8$ is a method to calculate the brittleness index and is frequently employed to predict brittleness using seismic data. This formula was deduced by Rickman *et al.* (2008) through massive experiments and production data. With $B_8$, it is considered that shale reservoirs with higher Young's modulus and lower Poisson's ratio are more brittle and respond better to hydraulic fracturing. For this formula, Young's modulus and Poisson's ratio are normalized first, and then their weighted values are added up to calculate the brittleness index.

(1) **Method for brittleness evaluation based on mineral composition**: The mineral composition and the proportions of brittle minerals are important factors affecting rock mechanical properties and directly determine the compressibility of rocks. Generally, the content of each mineral component of a rock can be determined through experiments and analysis of core samples of shale reservoirs. Based on this, the ternary diagram of mineral components can be plotted. Then, the compressibility of rocks can be evaluated using the contents of brittle minerals such as

**Table 5.1.** Formulas expressing the brittleness of shales.

| Definition of brittleness | Denotation |
|---|---|
| $B_1 = \frac{\varepsilon_{\theta 1}}{\varepsilon_{\text{tot}}}$ | Relationship between total stress and strain before breaking |
| $B_2 = \frac{W_{\theta 1}}{W_{\text{tot}}}$ | Relationship between total stress energy and strain before breaking |
| $B_3 = \frac{C_0 - T_0}{C_0 + T_0}$ | Ratio of the difference between compressive strength ($C_0$) and tensile strength ($T_0$) to the sum of $C_0$ and $T_0$ |
| $B_4 = \sin\varphi$ | $\varphi$ is the friction angle, measured when the normalized crushing envelope is 0 |
| $B_5 = \frac{T_{\max} - T_{\text{res}}}{T_{\max}}$ | Ratio of the difference between the maximum strength ($T_{\max}$) and residual strength ($T_{\text{res}}$) to the maximum strength |
| $B_6 = \left\lvert \frac{\varepsilon_f^p - \varepsilon_c^p}{\varepsilon_c^p} \right\rvert$ | $\varepsilon_f^p$ and $\varepsilon_c^p$ represent the plastic strain of damage and the plastic strain beyond a certain strain level, respectively |
| $B_7 = \text{OCR}^b$ | OCR means over-compaction ratio, which is the ratio of historical maximum vertical effective stress to current effective vertical stress |
| $B_8 = \frac{1}{2}\left(\frac{E_{\text{dym}[M_{\text{pst}}]}(0.8-\emptyset)-1}{8-1} E_{\text{dym}} + \frac{V_{\text{dym}-0.4}}{0.15-0.4}\right) \cdot 100$ | $E_{\text{dym}}$ and $V_{\text{dym}}$ refer to dynamic Young's modulus and Poisson's ratio, respectively |
| $B_9 = \frac{C_{\text{qtz}}}{C_{\text{qtz}} + C_{cl} + C_{\text{carb}}}$ | $C_{\text{qtz}}$, $C_{cl}$, and $C_{\text{carb}}$ refer to the content of quartz, clay, and carbonate, respectively |

quartz. The calculation formula is as follows:

$$\beta = \frac{C_{\text{quartz}}}{C_{\text{quartz}} + C_{\text{clay}} + C_{\text{carbonate}}} \tag{5.1}$$

where $\beta$ is the brittleness index or the proportion of brittle minerals, %; $C_{\text{quartz}}$ is the quartz content, %; $C_{\text{clay}}$ is the clay content, %; and $C_{\text{carbonate}}$ is the carbonate content, %.

(2) **Brittleness index method**: Young's modulus and Poisson's ratio are two basic rock mechanical parameters of rocks and are affected by lithology, physical properties, structure, and pore fluids of rocks. As indicated by the successful experience of shale gas development in North America, Young's modulus of shale gas reservoirs reflects the stiffness of rocks and directly determines the ability of the reservoirs to propagate fractures during hydraulic fracturing. Generally, a higher Young's modulus is associated with a higher ability of rocks to propagate fractures. Poisson's ratio reflects the ability of rocks to resist compression, representing the ability of rocks to resist crushing during loading. Generally, a smaller Poisson's ratio denotes that rocks are more prone to breaking after loading. Therefore, shale gas reservoirs with a higher Young's modulus and a lower Poisson's ratio have higher brittleness, are more prone to forming a complex fracture network during hydraulic fracturing, and are more favorable to producing high-yield shale gas.

Equation (5.2) are technical formulas for Young's modulus and Poisson's ratio according to the rock mechanical theory. According to this equation, the two parameters are generally inversely proportional. In other words, reservoirs with a higher Young's modulus have a lower Poisson's ratio and are more prone to fracturing and propagating fractures, and vice versa.

$$E = \frac{\mu(3\lambda + 2\mu)}{\lambda + \mu}, \quad \sigma = \frac{\lambda}{2(\lambda + \mu)} \tag{5.2}$$

where $E$ is Young's modulus, GPa; $\sigma$ is Poisson's ratio; $\lambda$ is the Lamé constant, GPa; $\mu$ is the shear modulus, GPa.

Through the study of the Barnett Shale in the Fort Worth Basin, American geophysicists proposed the following formula to quantitatively characterize the brittleness index of shale reservoirs based on

Young's modulus and Poisson's ratio:

$$\mathrm{BI} - \frac{1}{2}\left(E_{\mathrm{brt}} + \sigma_{\mathrm{brit}}\right) \tag{5.3}$$

$$E_{\mathrm{brt}} = \frac{E - E_{\mathrm{min}}}{E_{\mathrm{max}} - E_{\mathrm{min}}} \tag{5.4}$$

$$\sigma_{\mathrm{brt}} = \frac{\sigma - \sigma_{\mathrm{min}}}{\sigma_{\mathrm{max}} - \sigma_{\mathrm{min}}} \tag{5.5}$$

where $E$ is the static Young modulus; $E_{\mathrm{brit}}$ is the normalized Young modulus; $\sigma_{\mathrm{brit}}$ is the normalized Poisson ratio; $E_{\mathrm{min}}$ and $E_{\mathrm{max}}$ are the minimum and maximum values of static Young modulus, respectively, MPa; $\sigma$ is the static Poisson ratio; $\sigma_{\mathrm{max}}$ and $\sigma_{\mathrm{min}}$ are the maximum and minimum values of static Poisson ratio, respectively; and BI is the brittleness index of shale gas reservoirs, %.

A higher brittleness index calculated from Equation (5.3) corresponds to higher brittleness of the shale gas reservoirs. Therefore, the shale gas reservoirs with a higher brittleness index are more prone to fracturing and propagating fractures during hydraulic fracturing.

The rock mechanical properties directly determine the brittleness of shale gas reservoirs. Young's modulus and Poisson's ratio are usually used as the criteria for evaluating the brittleness of hydrocarbon reservoirs in the development of unconventional oil and gas. A higher Young's modulus means that the reservoirs have higher rigidity and hardness, and a lower Poisson's ratio means that the reservoirs have poorer expansibility. Therefore, a higher Young modulus and a lower Poisson ratio correspond to higher brittleness of shale reservoirs. There is a positive correlation between the brittle mineral content and brittleness of shales. However, it has been neglected that the brittleness changes even with the same mineral composition. With the same brittle mineral content, the brittleness will decrease under high-temperature and high-pressure conditions. During the shale gas exploration and development in North America, the spatial distribution of reservoir brittleness is mainly predicted using 3D seismic data. First, the data volumes of P-wave velocity, S-wave velocity, and density are obtained through pre-stack inversion. Then, the data volumes of Young's modulus and Poisson's ratio are obtained using basic petrophysical formulas. Finally, the brittleness index data volume is obtained using the formula presented earlier. Therefore, this study considers that the brittleness index is the most direct means to characterize the compressibility of strata and can be calculated solely using geophysical parameters, and that the brittleness index method is more direct and accurate than the brittleness evaluation method based on mineral composition.

### 5.1.2. *Spatial distribution of brittleness*

P-wave velocity, S-wave velocity, and density can be directly derived by full gather-based simultaneous pre-stack inversion, and then they can be used to calculate Young's modulus and Poisson's ratio. Figure 5.1 is a profile of Young's modulus, Poisson's ratio, and brittleness index obtained by pre-stack inversion. The reservoirs at the bottom of the Longmaxi Formation have a high Young modulus and low Poisson ratio, which are more prominent in the Long-$I_1$ Submember. The spatial distribution map of the brittleness index obtained from inversion shows that the Long-$I_1$ Submember generally has a high brittleness index, above 50%, with high lateral continuity. Therefore, the Long-$I_1$ Submember is the most favorable interval for hydraulic fracturing (Figure 5.1).

Figures 5.2–5.4 show the average distribution of Young's modulus, Poisson's ratio, and brittleness index of the Long-$I_1$ Submember in the Changning area. The main part of the Changning area has a high brittleness index, high Young's modulus, and low Poisson's ratio, indicating a sweet spot area favorable to hydraulic fracturing overall.

## 5.2. Seismic prediction of natural fractures

The development degree of natural fractures is an important factor controlling the accumulation of shale gas sweet spots and is also a key element of sweet spot prediction. Natural fractures play an important role in the following three aspects of shale gas exploration and development: (1) they provide a large amount of storage space; (2) they connect beddings to matrix pores, thus greatly improving the physical properties of shale gas reservoirs, enhancing their seepage capacity and permeability, and provide seepage pathways for the shale gas to migrate toward wellbores and for the adsorbed gas to be

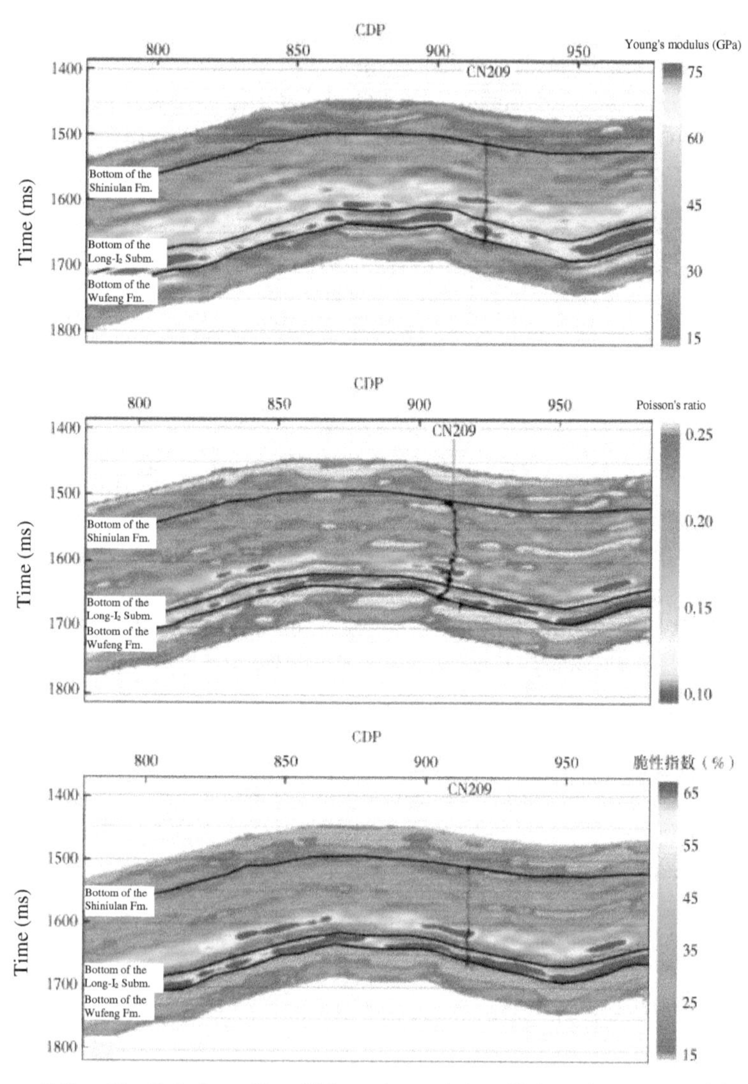


**Figure 5.1.** Predicted profile of Young's modulus, Poisson's ratio, and brittleness index along XLine591 passing through sell CN209.
*Note*: Fm. refers to Formation, and Subm. refers to Submember.

desorbed; (3) as mechanically weak surfaces in the rock, they facilitate the formation of induced large and complex fracture networks by hydraulic fracturing. Therefore, fracture prediction is a key technique in shale gas exploration and development.

**Figure 5.2.** Map showing the average distribution of Young's modulus of the Long-$I_1$ Submember.

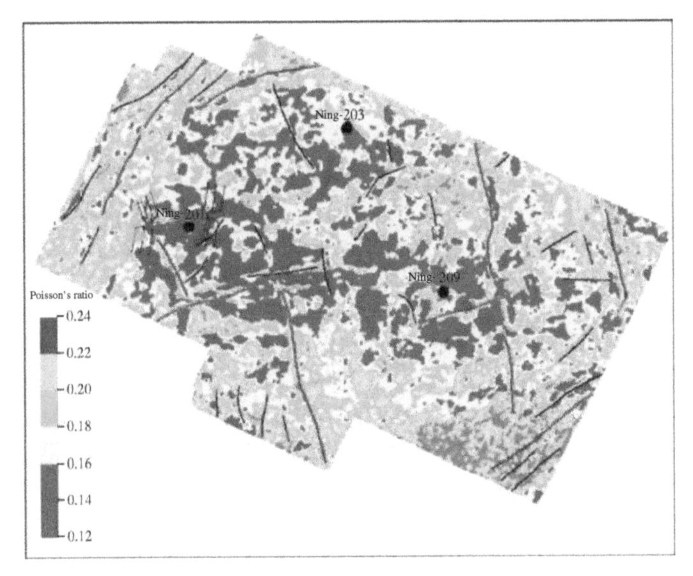


**Figure 5.3.** Map showing the average distribution of Poisson's ratio of the Long-$I_1$ Submember.

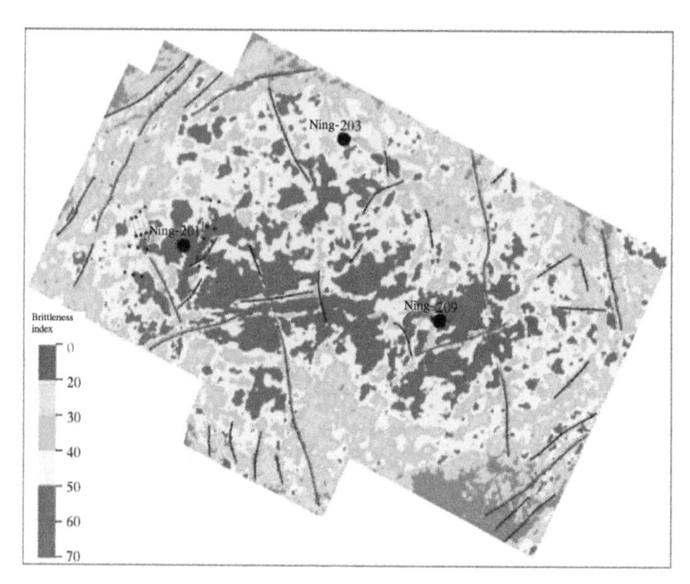


**Figure 5.4.** Map showing the average distribution of the brittleness index of the Long-$I_1$ Submember.

### 5.2.1. *Methods and principles*

The common geophysical techniques for the prediction and evaluation of fractures include log and seismic techniques. Log evaluation of fractures mainly depends on image log technology. There are many seismic techniques for fracture prediction, and the most common are based on seismic attributes such as curvature, coherence, variance, and ant tracking, or seismic anisotropy, and converted S-wave splitting methods.

(1) Fracture prediction techniques based on seismic attributes including seismic coherence, curvature, and variance. Curvature is commonly applied to represent the degree of bending of a curve or a surface, and higher curvature corresponds to a higher degree of bending. Most fractures in hydrocarbon reservoirs are related to tectonic stress and tectonic deformation. The more severe the tectonic deformation, the larger the networks of faults and fractures created. Curvature can therefore represent the degree of tectonic deformation of a stratum. Higher curvature means a higher deformation degree of a rock layer, where fractures are

more prone to form. Therefore, fractures are frequently indirectly predicted by calculating the curvature of strata seismically, and a higher curvature means a stronger possibility of the presence of fractures.

The waveform difference between seismic traces can be used to predict fracture presence using the coherence and variance attributes. The basic principle is as follows: when pores, fractures, and faults are present in rock layers, the time or amplitude of a reflected seismic wave will change, and the variation is represented by the distortion of the seismic waveform in terms of seismic signals. Once the coherence or variance between seismic traces is determined, the discontinuous distribution of seismic signals, which is closely related to the distribution of subsurface faults, can be obtained. Fractures are mostly present where the coherence or variance changes abruptly. The comprehensive application of these methods, combined with geological knowledge, can be used to qualitatively or semi-quantitatively predict the scale and orientation of fractures.

(2) Fracture prediction technique based on seismic anisotropy. Numerous scholars have conducted in-depth research and believe that all sedimentary strata show weak anisotropy on the seismic wave scale, which results from the directional arrangement of rock particles in the layers or the presence of fractures. Shale gas reservoirs have strong anisotropy because of the occurrence of bedding and fractures. Given this, a series of pre-stack azimuthal fracture prediction techniques have been developed, including the amplitude and velocity variations with azimuth–velocity (AVAZ-V) technique and the amplitude *vs.* offset/amplitude variation with azimuth (AVO/AVAZ) analysis technique.

(i) **AVAZ-V analysis technique**: The presence of beddings or fractures will result in anisotropy, which causes seismic velocity to change with changing azimuth. The strata with no fractures are isotropic, and their seismic wave velocity does not change with the azimuth. In this case, the plot of the seismic velocity versus azimuth is circular. On the contrary, layers with fractures are anisotropic, and the seismic wave velocity changes with the azimuth. In this case, the plot of the seismic velocity versus azimuth has an elliptic shape, whose long axis denotes a high stacking velocity and the direction parallel to fractures, and the short axis represents a low stacking velocity and the direction perpendicular to fractures. The flattening of the ellipse represents the fracture density. Since there is a direct correlation between the travel time and the seismic wave velocity, the direction and density of fractures can be predicted by analyzing the variation of the travel time with the azimuth.

(ii) **AVO/AVAZ analysis technique**: The presence of fractures in shale reservoirs leads to the creation of anisotropic reservoirs, and the induced seismic response has another characteristic, that is, the amplitude of the reflected waves changes with the azimuth. The strata without fractures are isotropic, and the amplitude of reflected waves and AVO gradient do not change with the azimuth, and vice versa. Accordingly, the density and orientation of fractures can be predicted by analyzing the amplitude variation of the reflected waves or the AVO gradient with the azimuth. Generally, reservoirs have the maximum anisotropy in the direction with the maximum amplitude of reflected waves and the maximum AVO gradient. This direction should be perpendicular to the fracture strike, and there should be a direct correlation between the AVO gradient and the fracture density.

The anisotropy-based fracture detection technique is accurate and quantitative. However, this technique requires high-quality original seismic data. Preliminary tests revealed that the seismic data of the study area are not suitable for anisotropy-based fracture detection, since they are affected by surface and underground conditions that limit their quality.

### 5.2.2. *Two-dimensional distribution characteristics of fractures*

Based on the geological characteristics of the study area, including the relationship between gas reservoirs and fractures as well as the comparison of multiple experimental methods, we concluded that the fractures in the study area are mainly caused by tectonic deformation and are closely related to structures. The curvature attribute accurately reflects small faults and fracture zones in this area. Therefore, this study finally proposed using the curvature technique to conduct seismic prediction of the density and lateral distribution of fractures in the Longmaxi Formation.

The relationship between curvature attributes and fractures is due to the stress changes when the strata fold, bend, or deform. The 3D curvature technique can directly extract the curvature of seismic events from post-stack data volume and then derive curvature attributes through further calculation of the curvature. Therefore, this technique can effectively describe the deformation degree of a strata, the three-dimensional (3D) and two-dimensional (2D) distribution of the *in-situ* stress field, and fracture density. This technology has been widely used for seismic-based fracture prediction in complex reservoirs due to its reliability and accuracy.

The curvature of a curve can be defined as follows:

$$K = \left|\frac{d\varphi}{ds}\right| = \left|\frac{y''}{(1+y'^2)^{3/2}}\right| \tag{5.6}$$

where $K$ is the curvature of the curve at a point; $\varphi$ is the direction angle corresponding to the tangent line at a point of the curve; s is the arc length corresponding to the direction angle; $y'$ and $y''$ are the first and second derivatives of the curve, respectively. A higher degree of bending of a curve corresponds to higher curvature.

According to Equation (5.6), the curvature of a curve is closely related to its second derivative, and there is no difference between positive or negative curvature. However, the positive and negative values have different geological meanings in seismic exploration: (1) a zero curvature denotes a flat layer; (2) positive curvature suggests a positive structure, such as an anticline or uplift; (3) negative curvature indicates a negative structure, such as a syncline or sag.

Since an actual formation interface is a 3D curved surface, previous researchers extended the definition of curvature from 2D curves to 3D curved surfaces and pointed out that a curve can be determined between the formation interface and any plane cutting it. According to the calculation equation of curvature, the curvature at any point of the curve can be calculated. Since there is an infinite number of planes that cut a 3D surface at any point, there are numerous curvatures at that point. In practical application, the most significant curvatures are those determined by planes that vertically intersect the formation interface, which are useful for practical production after being combined in different ways. The curvatures that are commonly used to describe the formation attitude and the fracture density include inclination curvature, strike curvature, the Gaussian curvature, maximum curvature, minimum curvature, maximum positive curvature, and minimum negative curvature.

In general, a 3D curved surface corresponding to a formation can be approached using the following binary quadratic equation with space coordinates as variables:

$$z(x, y) = ax^2 + by^2 + cxy + dx + ey + f \tag{5.7}$$

The attributes of curvatures such as the Gaussian curvature can be obtained based on coefficients $a$, $b$, $c$, $d$, $e$, and $f$ in Equation (5.7).

According to the abovementioned principle, this study predicted the fractures in the shale reservoirs of the Longmaxi Formation (Long-$I_1$ Submember) in the study area using the interlayer attributes extracted from the bottom boundary of the Longmaxi Formation to 15 m above the the bottom of the Long-$I_2$ Submember. The resulting fracture prediction map of the Long-$I_1$ Submember is shown in Figure 5.5.

The red zones in the fracture prediction map are areas with faults. As informed by superimposed faults in the map, areas with faults are bound to have high curvature. The faults in these areas affect the accumulation and preservation of shale gas and thus are detrimental to the development of shale gas. By contrast, the areas with high curvature but no faults indicate the occurrence of fracture zones. These areas have ideal conditions for shale gas preservation and superior physical properties of reservoirs and are favorable areas for shale gas exploration and development.

As shown by the fracture prediction results, fractures are developed in the main portion of the study area and show considerable consistency. The main fracture set has a nearly NW-SE strike and a subordinate fracture set has a nearly NE-SW strike. Practical production has revealed that the reservoirs have high gas content. They are prone to crack in late fracturing, thus favoring the fracturing of shale reservoirs and the release of shale gas.

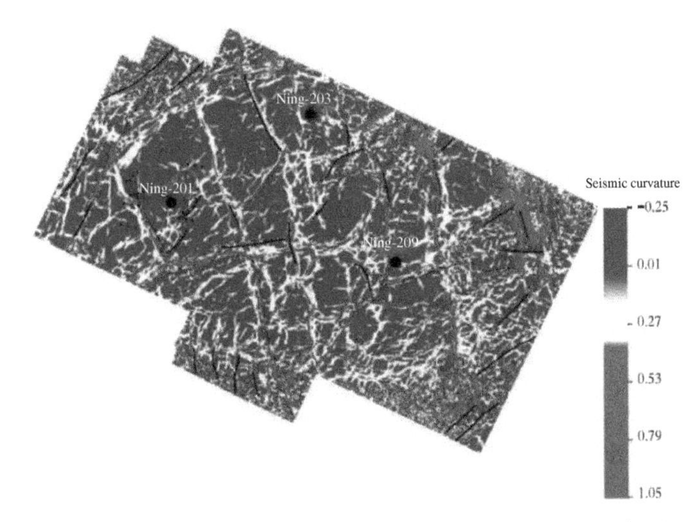


**Figure 5.5.** The fracture prediction map (Long-$I_1$ Submember) of Longmaxi Formation reservoir.

## 5.3. Seismic prediction of pore pressure

### 5.3.1. *Geological significance for oil and gas and classification of pore pressure*

Formation pressure, also known as pore pressure, refers to the pressure borne by fluids in the formation pores. The formation pressure of a reservoir reflects the energy of oil and gas zones and is the main force driving oil and gas migration. The prediction of formation pressure is greatly significant for oil and gas exploration and development. In the oil and gas exploration phase, anomalous formation pressure, especially anomalously high formation pressure, tends to indicate oil and gas accumulation. The statistical analysis of the formation pressure of more than 160 well-known hydrocarbon reservoirs around the world shows that reservoirs with normal formation pressure account for 37.5% of the total reservoirs, those with high pressure or overpressure account for 47.7%, and those with low or ultra-low pressure account for 14.8%. Research on the distribution of pore pressure (especially anomalous pore pressure) in hydrocarbon reservoirs facilitate the rapid location of hydrocarbon reservoirs. In the drilling stage, an accurate prediction of formation pressure before drilling can ensure safe and fast drilling, provide an important basis for determining the specific gravity of drilling fluids and casing programs, and provide important parameters for determining drilling equipment types and completion methods. In the field development phase, the prediction and the distribution pattern research of formation pressure can help locate new hydrocarbon reservoirs, understand the underground energy distribution, and control the change in reservoir energy, aiming to make rational use of the reservoir energy and maximize the recovery rate.

As for shale gas exploration and development in North America, which has stable geological conditions and favorable preservation conditions overall, it is generally considered that reservoir overpressure is not that essential for shale gas accumulation there. Moreover, there is no significant overpressure characteristic in North America, where most shale gas fields have normal pressure and even low-pressure reservoirs can be commercially exploited. Therefore, reservoir overpressure is not taken as a key evaluation indicator in the shale gas sweet spot prediction in North America. By contrast, South China has a significantly different geological background of shale gas from North America. Specifically, the main shale gas reservoirs in North America were formed in stable craton basins, while those in South Sichuan were formed in complex superimposed basins a very long time ago and underwent multiple intense tectonic movements. Therefore, the most significant difference in shale gas geological conditions between South Sichuan and North America lies in the shale maturity and the intensity of tectonic movements. The shale gas in South Sichuan is characterized by a higher maturity, a longer geological history, stronger compaction, lower reservoir porosity, more phases of intense tectonic movements, more developed fault networks, and poorer preservation conditions. The exploration and development experience also shows that the shale gas fields that have been developed in China so far are all high-pressure or overpressure gas reservoirs with obvious overpressure characteristics. Moreover, the shale gas fields in North America include both high-pressure and low-pressure gas reservoirs, as shown in Figure 5.6, while for marine shale gas in South China, high reservoir pressure serves as both a sign of favorable preservation conditions and a prerequisite for the formation of sweet spots.

Formation pressure is the pressure acting on the fluids in the pore space of strata. When the pore pressure is higher than the hydrostatic

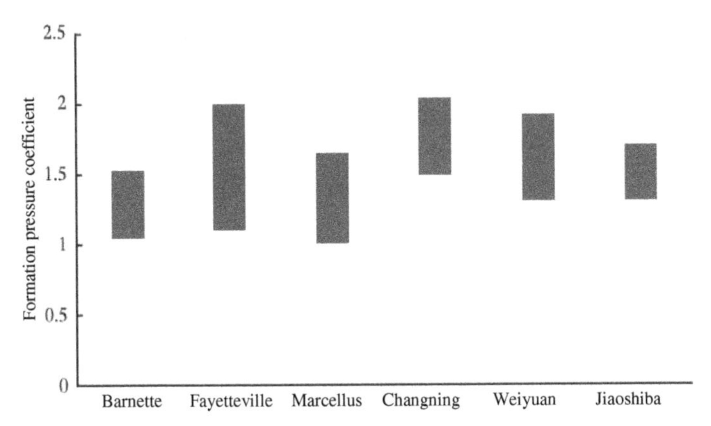


**Figure 5.6.** Distribution of formation pressure coefficient in major shale gas fields at home and abroad.

**Table 5.2.** Classification criteria for formation pressure.

| No. | Pressure class | Pressure gradient (MPa) | Pressure coefficient |
|---|---|---|---|
| 1 | Low pressure | ≤10.0 | ≤0.8 |
| 2 | Normal pressure | 10.0–12.7 | 0.8–1.27 |
| 3 | Pressure transition zone | 12.7–15.0 | 1.27–1.5 |
| 4 | Overpressure | 15.0–17.3 | 1.5–1.73 |
| 5 | Ultrahigh pressure | ≥17.3 | ≥1.73 |

pressure, the pore fluids are confined in the pores and partially bear the overburden pressure. In this case, the pore pressure is referred to as formation overpressure. Otherwise, it is known as anomalous low pressure. In addition, the formation pressure equal to or close to the hydrostatic pressure is referred to as normal pressure. The pressure coefficient, which is used to measure whether the formation pressure is normal, is the ratio of the measured formation pressure to the hydrostatic pressure at the same depth. The classification criteria for the pressure coefficient are shown in Table 5.2. An anomalous formation pressure mostly results from the combined action of multiple factors, including (i) thermal pressurization of fluids; (ii) denudation; (iii) faulting and lithologic sealing; (iv) piercing; (v) buoyancy, and (vi) diagenetic evolution of clay minerals.

### 5.3.2. *Pore pressure characteristics of shale gas reservoirs*

As revealed by the experience in marine shale gas exploration and development in South China, formation overpressure is both the main factor controlling the presence of commercially favorable shale gas reservoirs and the sign of good preservation conditions. The accurate prediction of formation pressure before drilling can ensure safe drilling and is also very important for the prediction of sweet spots and the optimal deployment of horizontal wells.

Figure 5.7 shows the results of the pressure tests and production analysis of eight wells drilled in the study area and its adjacent blocks. There is a significant positive correlation between the formation pressure coefficient and the test production. A higher formation pressure coefficient corresponds to a higher test production and greater potential for shale gas development. The quantitative relationship between them can be expressed using an exponential function, with a correlation coefficient greater than 96%.

Figures 5.8 and 5.9 show the analysis results of the relationships of formation pressure and formation pressure coefficient with the burial depth. Figure 5.8 shows the results of pressure tests of eight wells, with the $y$-axis denoting the pressure test depth. According to this figure, there is a close positive correlation between the formation

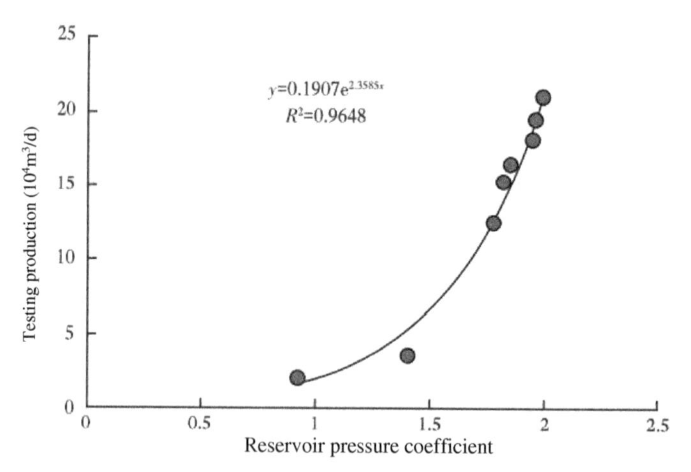


**Figure 5.7.** Relationship between formation pressure coefficient and test production.

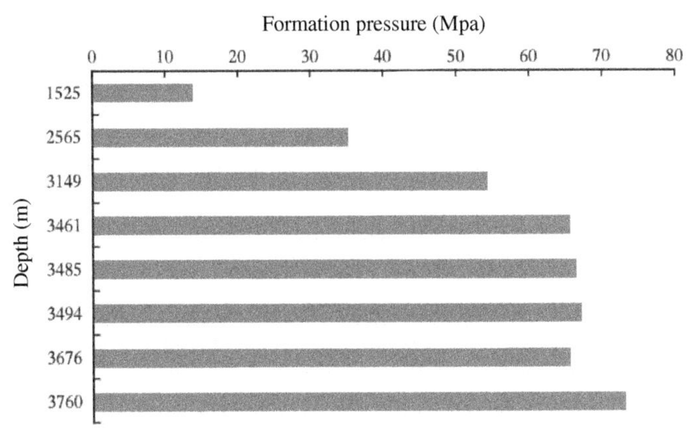


**Figure 5.8.** Distribution of formation pressure of the target intervals of eight wells.

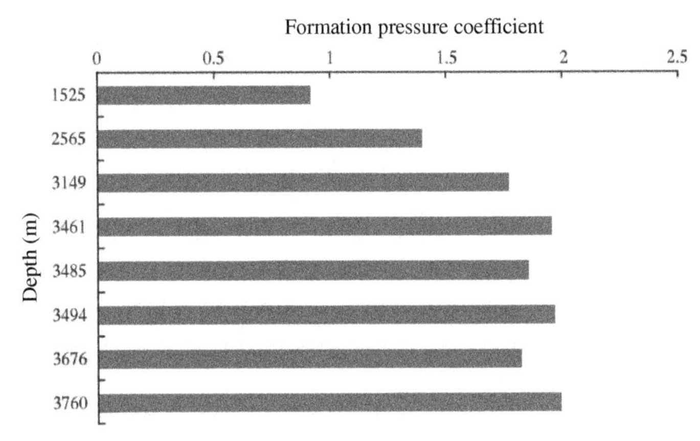


**Figure 5.9.** Distribution of formation pressure coefficient of the target intervals of eight wells.

pressure and the burial depth, that is, a greater burial depth corresponds to higher formation pressure. Moreover, the target strata of most wells have formation pressure of 50–70 MPa. Figure 5.9 shows the relationship between the formation pressure coefficient and the depth of the eight wells, with the $y$-axis denoting the pressure test depth and the $x$-axis representing the formation pressure coefficient. This figure indicates that there is a close positive correlation between the formation pressure coefficient and the burial depth and that the formation pressure coefficient is always above 1.5 at a burial depth over 2,500 m, indicating significant formation overpressure. The target intervals of most wells have a formation pressure coefficient of 1.5–2.0.

This study analyzed all formation pressure prediction methods published so far. These methods can be divided into two types according to the calculation formulas they adopt. One type of methods is based on the normal compaction trend line (NCTL), and their prerequisite is to plot a NCTL from shallow to deep parts according to the formation lithology and porosity. They are mainly used for logging-based pressure prediction. The other type of methods are based on the principle of effective stress rather than NCTLs and determine the presence of anomalous formation pressure using formation velocity anomalies. These methods are mostly used for the seismic prediction of formation pressure.

### 5.3.3. *NCTL-based prediction methods of formation pressure*

The NCTL-based methods include the Brewer method, the equivalent depth method, the Eaton method, the improved Eaton method, and the Stoneley method. Table 5.3 shows the calculation models, supporting data, and the inventor of each method.

For these methods, accurate prediction results of formation pressure depend on NCTLs, especially for deeply buried layers, for the following reasons. The errors of shallow layers will gradually add up to the final pressure prediction result of the deep layers with an increase in the burial depth, forming a significant error. Moreover, a

**Table 5.3.** NCTL-based prediction methods of formation pressure.

| Method | Formulas |
|---|---|
| Brewer method | $P_{fb} = G_0 H_B + (G_W + G_0) H_A$ |
| Equivalent depth method | $P_A = S_A - (S_B - P_B)$ |
| Eaton method | $P_p = S - (S - P_n)(R/R_n)^2$ |
| | $P = S - (S - P_n)(R/R_n)^3$ |
| Improved Eaton method | $P = S - (S - P_n)$ |
| | $\{(\Delta t_{\mathrm{ma}} + (\Delta t_{\mathrm{ml}} - \Delta t_{\mathrm{ma}})e^{-C_{n^z}})/\Delta t\}^3$ |
| Stoneley method | $P_f = H \Delta t \frac{\Delta t}{\Delta t_n} C$ |

widely applicable method to develop NCTLs has not been found yet, and the NCTLs can only be built based on experience. In particular, for discontinuous sedimentary sequences, different sedimentary intervals correspond to different NCTLs, and it is nearly impossible to establish an accurate NCTL. These factors significantly reduce the prediction accuracy of these methods, which heavily rely on data availability and experience in the work areas. Furthermore, these methods have another major weakness, i.e., these methods can only be applied to predict the formation pressure at the well locations after drilling and to mature blocks with multiple wells.

### 5.3.4. *Prediction methods of formation pressure based on the principle of effective stress*

To address the shortcomings of the NCTL-based formation pressure prediction methods, many researchers have proposed methods relying on the principle of effective stress rather than NCTLs. These methods include the Fillippone method, the Bowers method, the Miller method, the Tau method, the Liu Z method, the Zhang method, the Eberhart-Phillips (E-P) model method, and the Bellotti method. Table 5.4 shows the calculation formulas and supporting data of each method.

The basic principle of effective stress-based methods is as follows. The rock matrix of reservoirs and pore fluids jointly bear the overburden pressure. The part of the overburden pressure borne by the rock matrix is equal but opposite to the effective stress. Under a fixed overburden pressure, lower effective stress means that pore fluids bear higher overburden pressure and the reservoirs have more significant overpressure. On the contrary, higher effective stress means that the rock matrix and pore fluids bear the major and minor portions of the overburden pressure, respectively. In this case, the reservoirs feature abnormally low pressure.

These methods indirectly predict pore pressure by calculating the overburden pressure and effective stress, without the need to develop NCTLs. In this manner, these methods eliminate errors arising from inaccurate NCTLs and are convenient to use. However, these methods involve many assumptions in the derivation process, and their prediction accuracy is largely limited by the conformity of the assumptions to the actual situation in the study area. In addition, these methods commonly rely on a pretty simple equation that

**Table 5.4.** Prediction methods of formation pressure based on the principle of effective stress.

| Method | Formula |
|---|---|
| Fillippone method | $P_f = \frac{V_{\max} - V_1}{V_{\max} - V_{\min}}$ |
| | $P_f = \frac{V_{\max} - V_1}{V_{\max} - V_{\text{mnp}}}$ |
| Bowers method | Deposit loading: $V_p = 5{,}000 + A\sigma^B$ |
| | Deposit unloading : $V_p = 5{,}000 + A[\sigma_{\max}(\sigma/\sigma_{\max})^{1/\mu}]^B$ |
| Miller method | When $Z < d_{\max}$: |
| | $P = S - \frac{1}{\lambda}\text{In}\left[\frac{\Delta t}{\Delta t_{\text{ml}}}\left(\frac{\Delta t_{\text{ml}} - \Delta t_{\text{ma}}}{\Delta t - \Delta t_{\text{ma}}}\right)\right]$ |
| | When $Z < d_{\max}$: |
| | $P = S - \frac{1}{\lambda}\text{In}\left[a\left(1 - \frac{\frac{1}{\Delta t} - \frac{1}{\Delta t_{\text{uo}}}}{\frac{1}{\Delta t_{\text{ma}}} - \frac{1}{\Delta t_{\text{ml}}}}\right)\right]$ |
| | $\frac{10^6}{\Delta t_{\text{uo}}} = V_{\text{ml}} + (V_m - V_{\text{ml}})\, e^{-\lambda\sigma} \max\left(\frac{1-\gamma}{\gamma}\right)$ |
| Tau method | $\sigma = A_s T^{B_s}$ |
| | $T = \frac{V_{\text{ma}}}{V_f}\frac{V_p - V_f}{V_{\text{ma}} - V_p}$ |
| Liu Zhen method | $P_f = \frac{\text{In}(V_i/V_{\max})}{\text{In}(V_{\text{mini}}/V_{\max})}\text{Pov}$ |
| Zhang method | $P = S - \frac{\sigma}{bZ}\left(\frac{b - C_n}{C_n}\text{In}\frac{\Delta t_{\text{ml}} - \Delta t_{\text{ms}}}{\Delta t_{\text{uo}} - \Delta t_{\text{ms}}}\right) + \text{In}\frac{\Delta t_{\text{ml}} - \Delta t_{\text{ms}}}{\Delta t_{\text{uo}} - \Delta t_{\text{ms}}}$ |
| E-P method | $V_p = 5.77 - 6.94\emptyset - 1.73\sqrt{V_{\text{sh}}} + 0.46(\sigma - e^{-16.7\sigma})$ |
| | $V_s = 3.70 - 4.94\emptyset - 1.57\sqrt{V_{\text{sh}}} + 0.36(\sigma - e^{-16.7\sigma})$ |
| Bellotti method | $P_f(Z_i) = \text{Pov}(Z_i)[(C_{\max}(Z_i) - C_i(Z_i)]/[C_{\max}(Z_i) - C_{\min}(Z_i)]$ |

includes only the P-wave velocity in the formation, while ignoring the fact that the S-wave velocity rather than its P-wave velocity can best reflect the changes in the effective stress. Many petrophysical studies have shown that the S-wave velocity is almost only related to the rock matrix stress, while the P-wave velocity is greatly affected by pore fluids. When only the P-wave velocity is used to predict the formation pressure using these methods, the fluid-induced reduction in velocity is prone to be considered the effect of rock matrix stress by mistake. As a result, the predicted formation pressure will be higher than the actual value.

### 5.3.5. *Improvement and innovation of the methods*

To overcome the limitations of the abovementioned methods, this study re-derived a prediction method for formation pressure based on rock mechanics after inheriting their advantages.

The principle of effective stress for porous reservoirs at any burial depth can be expressed as follows:

$$P_{\text{ov}} = \sigma + P_f \tag{5.8}$$

where $P_{\text{ov}}$ is the overburden pressure, MPa; $\sigma$ is the effective stress on rock matrix, MPa; $P_f$ is pressure borne by pore fluids, aka the formation pressure, MPa.

The formation pressure can be calculated as follows in the case of known overburden pressure and effective stress:

$$P_f = P_{\text{ov}} - \sigma \tag{5.9}$$

The overburden pressure, i.e. the gravity of rocks above the reservoir, can usually be treated as a linear function of the burial depth. To ensure the prediction accuracy of formation pressure, this study calculated the overburden pressure strictly according to its physical meaning. The calculation formula is as follows:

$$P_{\text{oV}} = \bar{p}gh \tag{5.10}$$

where $\bar{p}$ is the average density of the overlying strata, g/cm$^3$; $g$ is the acceleration of gravity, m/s; h is the burial depth of the reservoir, m.

For the calculation of effective stress, Domenko's rock mechanical experiments conducted in 1977 revealed that effective stress is closely related to the P-wave (or S-wave) velocity under laboratory conditions. The results of seismic petrophysical research further showed that shear waves do not propagate through fluids, and thus the S-wave velocity is only affected by the stress on the rock matrix. Therefore, the shear waves are used to characterize rock stress in this study. According to the generalized Hooke law:

$$K = \frac{\sigma}{\Delta V/V} \tag{5.11}$$

where $K$ is the bulk modulus of rock matrix, MPa; $\sigma$ is the effective stress, MPa; $\Delta V/V$ is the amount of compression per unit volume of rocks, aka volume strain.

The relationship between P-wave velocity, S-wave velocity, and rock mechanical parameters can be derived from the wave equation, as follows:

$$v_s = \sqrt{\frac{\mu}{\rho}} \tag{5.12}$$

$$v_p = \sqrt{\frac{K + \frac{4}{3}}{\rho}} \tag{5.13}$$

where $v_s$ is the S-wave velocity in rocks, m/s; $v_p$ is the P-wave velocity in rocks, m/s; $\mu$ is the shear modulus of rocks, MPa; $K$ is the bulk modulus of rocks, Mpa; and $p$ is the rock density, g/cm$^3$.

The effective stress calculation formula can be obtained from Equations (5.12), (5.13), and (5.11):

$$\sigma = \rho\left(V_p^2 - \frac{4}{3}V_s^2\right) \tag{5.14}$$

Generally, in a normal sedimentary basin, the horizontal strain is much smaller than the vertical strain during the compaction of a unit volume element. Therefore, the strain per unit volume ($\Delta V/V$) is approximately equal to the strain per unit thickness ($\Delta H/H$):

$$\sigma = p\left(V_p^2 - \frac{4}{3}V_s^2\right)\frac{\Delta H}{H} \tag{5.15}$$

Substituting Equations (5.15) and (5.10) into Equation (5.9) yields:

$$P_f = \bar{p}gh - p\left(V_p^2 - \frac{4}{3}V_s^2\right)\frac{\Delta H}{H} \tag{5.16}$$

The compression per unit thickness is equal to the gradient of porosity. The relationship between porosity and depth variation:

$$\phi = \phi_0 e^{ch} \tag{5.17}$$

where $\phi$ is the porosity at depth h, %; $\phi_0$ is the porosity of shales at zero depth, i.e. the critical porosity of shales, %; $c$ is a dimensionless constant; $h$ is the depth, m.

Thus, the amount of compression of rocks per unit thickness can be approximated by the gradient of porosity:

$$\frac{\Delta H}{H} = \frac{\phi - \phi_0}{H} = \frac{\phi_0(1 - e^{ch})}{H} \tag{5.18}$$

The new formation pressure calculation formula can finally be obtained, as follows:

$$P_f = \bar{p}gh - p\left(V_p^2 - \frac{4}{3}V_s^2\right)\frac{\phi_0(1 - e^{ch})}{H} \tag{5.19}$$

Equation (5.19) shows that $c$ is the formation compaction coefficient, which can be obtained by fitting the porosity data of wells. The formation pressure is a function of P-wave velocity, S-wave velocity, density, depth, and critical porosity, out of which the first three can be obtained through pre-stack inversion, and the critical porosity can be obtained by petrophysical modeling of shale reservoirs in the study area. In this equation, the P-wave velocity, S-wave velocity, and density are taken as variables of pore pressure, and in particular, S-wave velocity is introduced in the calculation of the effective compressive stress. Since shear waves do not propagate in fluids, their variations only depend on the composition of the rock matrix and effective stress. Therefore, the introduction of S-wave velocity can improve the calculation accuracy of effective stress, which in turn improves the prediction accuracy of formation pressure.

To investigate the relationship between the formation pressure and acoustic velocity in the Changning block, this study prepared the cross plot of the formation pressure of the target intervals and the P-wave velocity at the pressure measuring points in the Changning block and its adjacent areas, as shown in Figure 5.10. According to this figure, the P-wave velocity decreases with increasing formation pressure, and so does the S-wave velocity. However, the distribution of points on the cross plot shows that there is no significant quantitative relationship between formation pressure and acoustic velocity. Furthermore, gas in the reservoirs will also decrease acoustic velocity. Therefore, it will inevitably cause low prediction accuracy when predicting the formation pressure simply using only P-wave velocity or S-wave velocity.

The abovementioned analysis and formula derivation show that the determination of critical porosity plays an important role in

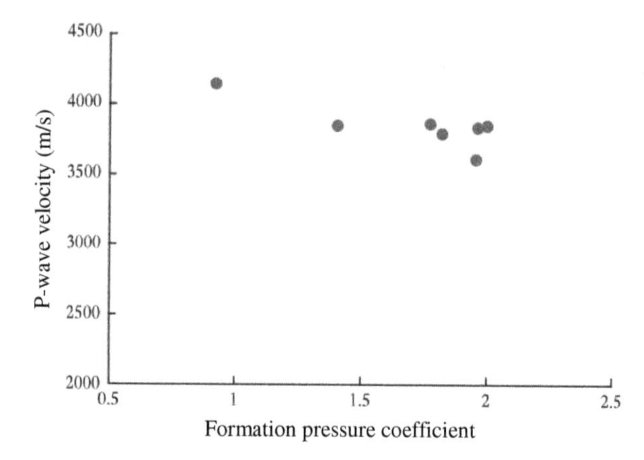


**Figure 5.10.** Cross plot of formation pressure and P-wave velocity.

improving the prediction accuracy of formation pressure. The critical porosity of rocks corresponds to the original sedimentary state of rocks immediately before compaction and diagenesis.

Numerous petrophysical experiments have shown that most rocks with stable lithology and mineral composition have a fixed critical porosity. The porosity of rocks below their critical porosity indicates that rocks have been compacted and undergone diagenesis, the mineral particles in the rocks are in contact with each other, and effective stress is transmitted through the solid parts of the rocks. When the porosity of rocks is above their critical porosity, the mineral particles in the rocks are substantially apart from each other, and effective stress is mainly transmitted through pore fluids. Many rock mechanical formulas and principles in reservoir rock mechanics involve critical porosity.

Using critical porosity, the relationship between the elastic modulus of the rock matrix and solid matrix:

$$K_{\mathrm{dry}} = K_{\mathrm{ma}}\left(1 - \frac{\phi}{\phi_c}\right) \tag{5.20}$$

$$\mu_{\mathrm{ma}} = \left(1 - \frac{\phi}{\phi_c}\right) \tag{5.21}$$

where $K_{\mathrm{dry}}$, $K_{\mathrm{ma}}$, $\mu_{\mathrm{mdrya}}$, and $\mu_{\mathrm{ma}}$ are the bulk modulus of the rock matrix, the bulk modulus of the solid matrix, the shear modulus of the rock matrix, and shear modulus of the solid matrix, respectively.

Equations (5.20) and (5.21) are often used to calculate the rock moduli in the case of known porosity and critical porosity of rocks. In this study, the rock moduli and the porosity were calculated using the data on logs and mineral composition. Specifically, the porosity $\phi$ was calculated using a multi-mineral optimization algorithm, and the bulk modulus $K_{\text{ma}}$ and shear modulus $\mu_{\text{ma}}$ of the solid matrix were calculated based on the bulk modulus and shear modulus of various mineral components using the Voigt–Reuss–Hill model. Moreover, the bulk modulus and shear modulus of various mineral components were calculated based on the P-wave velocity, S-wave velocity, and density of various components using the basic petrophysical formulas. In addition, $K_{\text{dry}}$ and $\mu_{\text{dry}}$ can be calculated using the Krief model on the premise that porosity, $K_{\text{ma}}$, and $\mu_{\text{ma}}$ are known. Finally, the critical porosity $\phi_c$ can be calculated based on these known variables using Equations (5.20) and (5.21). The marine shales in the study area have consistent lithology and mineral composition. Table 5.5 shows the mineral composition and corresponding density, P-wave, and S-wave velocities, which are sourced from the log data of the target intervals.

According to the curves of the porosity, bulk modulus, and shear modulus calculated from log data, the critical porosity of shales in the Changnning block can be determined by substituting the average bulk modulus, shear modulus, and porosity of the target intervals into Equation (5.21).

After accurately calculating the critical porosity, the prediction accuracy of formation pressure depends on the accuracy of the

**Table 5.5.** Mineral composition and physical parameters of rocks in the Longmaxi Formation.

| Mineral component | Content (%) | Density (g/m$^3$) | P-wave velocity (m/s) | S-wave velocity (m/s) |
|---|---|---|---|---|
| Quartz | 55.8 | 2.65 | 6,050 | 4,100 |
| Plagioclase | 6.0 | 2.62 | 6,220 | 3,580 |
| Calcite | 9.5 | 2.71 | 6,400 | 3,350 |
| Dolomite | 8.2 | 2.87 | 7,100 | 4,100 |
| Pyrite | 2.5 | 5.00 | 7,700 | 4,900 |
| Clay | 12.5 | 2.12 | 2,500 | 1,130 |
| Kerogen | 5.5 | 1.30 | 1,910 | 1,050 |

P-wave velocity, S-wave velocity, and density. In this study, the P-wave velocities, S-wave velocities, and densities of formations were obtained using the simultaneous pre-stack inversion technique. The optimal elastic parameter distribution model built based on the Bayes principle of this technique fully utilizes the rich seismic reflection information of reservoirs contained in pre-stack gathers. Therefore, this technique ensures the consistency of the inversion results of P-wave impedance, S-wave impedance, and density and can obtain accurate data volumes of P-wave velocity, S-wave velocity, and density, thus guaranteeing the optimal prediction of formation pressure.

Based on the data volumes of P-wave velocity, S-wave velocity, and density obtained through the simultaneous pre-stack inversion, the data volumes of formation pressure and formation pressure coefficient can be derived by applying the formation pressure prediction method presented in this study. Figure 5.11 shows the profiles of predicted formation pressure and pressure coefficient, with arrows pointing to the target intervals. The predicted results show that the formation pressure at the same horizon increases gradually as the

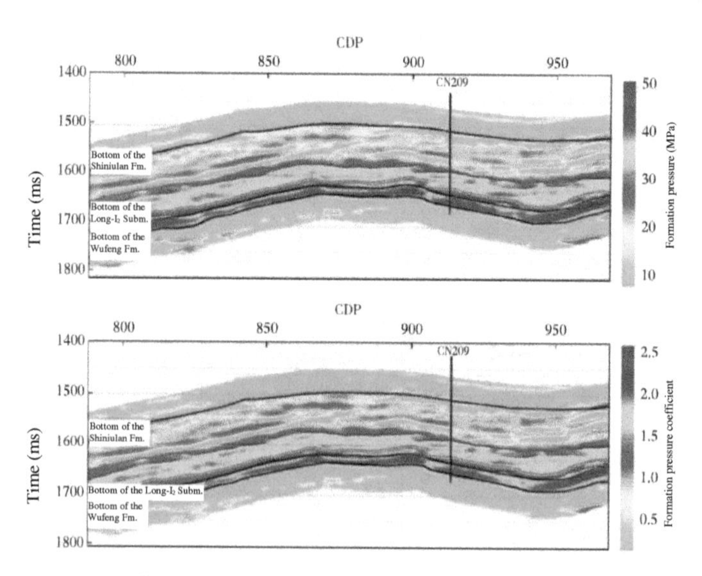


**Figure 5.11.** Profiles of predicted formation pressure and pressure coefficient. *Note*: Fm. refers to Formation.

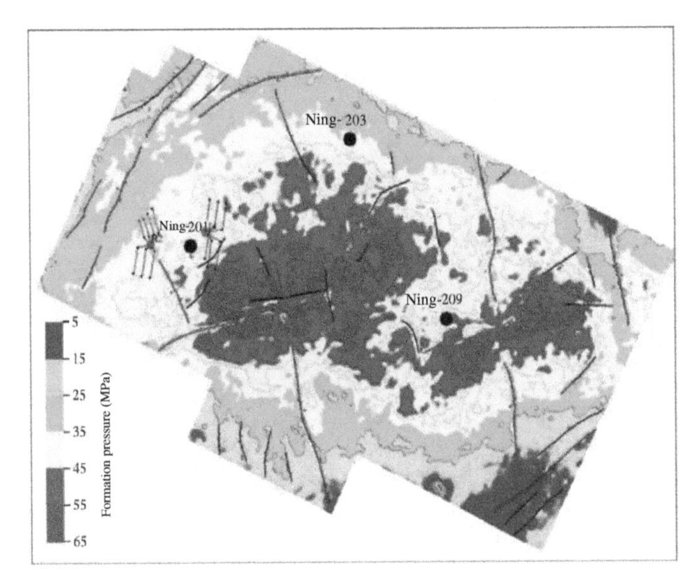


**Figure 5.12.** Map showing the planar distribution of predicted average formation pressure of the Changning area.

depth increases from left to right in the profiles. Vertically, the target intervals have high formation pressure overall, with significant high-pressure anomalies.

Figure 5.12 shows the map of predicted average formation pressure at the target intervals, which was calculated within the time window limited by the top and bottom of the target intervals based on the predicted data volume of formation pressure. Warm colors in the figure represent zones with high formation pressure, while cold colors represent those with low formation pressure. The formation pressure gradually increases from the northwest (20–30 MPa) to the southeast (40–70 MPa) in the study area. The higher formation pressure in the southeast indicates favorable gas-bearing properties and preservation conditions. Therefore, zones in the southeast are favorable for shale gas development.

According to the definition of pressure coefficient, dividing the data volume of the formation pressure by the hydrostatic pressure data volume yielded the data volume of the formation pressure coefficient. Based on this, the 2D and 3D distributions of the formation pressure coefficient can be obtained. The distribution of the

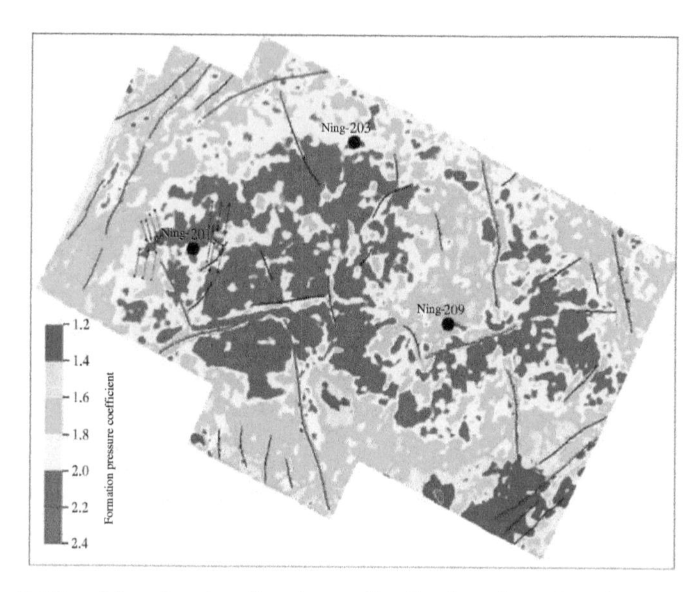


**Figure 5.13.** Map showing the planar distribution of predicted average formation pressure coefficient of the Changning area.

formation pressure coefficient is shown in Figure 5.13. In this figure, warm red and purple colors represent zones with high pressure and overpressure, respectively, with formation pressure coefficients over 2. These zones are the sweet spots that are most favorable for shale gas development and, thus, are recommended to be preferentially developed. Cold blue and green colors represent normal pressure zones, with pressure coefficients of approximately or below 1. These zones are considered to have low gas content and poor preservation conditions, and it is not recommended to deploy horizontal wells in these zones.

## 5.4. Seismic prediction of the *in-situ* stress

*In-situ* stress is the natural stress without engineering disturbance in the Earth's crust and is also known as the initial stress of rock masses, absolute stress, or stress of protolith. In a broad sense, it also refers to the stress in the Earth's interior. It includes stresses caused by factors including geothermal energy, gravity, and changes in the Earth's rotational speed.

Generally, the stress state varies from point to point within the Earth's crust and increases linearly with the burial depth. Different structural situations and geographical locations correspond to different stress gradients. The total spatial distribution of the stress states in the Earth's crust is called the crustal stress field. The crustal stress field related to geological tectonic movement, known as the tectonic stress field, usually refers to the crustal stress field that causes tectonic movements. According to geomechanics, it is the stress activities in the Earth's crust that enable the Earth's crust to overcome its own internal resistance, deform, and move, and all deformations in the Earth's crust, such as folds and faults, are the results of *in-situ* stress.

The tectonic stress field in geomechanics refers to the *in-situ* stress field that forms the tectonic style, including the distribution areas of the tectonic style and the stress distribution within it when the tectonic style was formed. The existing or active *in-situ* stress field is called the present tectonic stress field. To study the tectonic stress field, it is necessary to investigate the tectonic deformation of rocks and the uplifting and subsidence of regions during the most recent period (especially since the Quaternary) and to directly measure the current *in-situ* stress using appropriate instruments and methods. During the measurement of *in-situ* stress, the measurement points should be arranged according to the active tectonic systems, active tectonic zones (e.g., earthquake zones), and the requirements of major engineering construction, as well as the requirements for cooperation with the corresponding geological work.

### 5.4.1. *Significance of in-situ stress prediction*

The state of the *in-situ* stress is of great significance to engineering fields such as earthquake forecast, the assessment of regional crustal stability, the stability of boreholes in oil and gas fields, the storage of nuclear waste, and research on rock bursts, coal and gas outbursts, and geodynamics.

Shale gas reservoirs have lower porosity and permeability than conventional gas reservoirs such as carbonate rocks and sandstones. The key to large-scale exploitation of shale gas reservoirs lies in the application of horizontal drilling and hydraulic fracturing techniques. A large number of induced fracture networks and their connections with natural fractures are the keys to the large-scale efficient development of shale gas reservoirs. Many physical and digital simulation experiments and shale gas development practices have shown that when the horizontal well drilling direction is parallel to the direction of the minimum horizontal principal stress, multiple groups of complex fracture networks perpendicular to the well trajectory direction can be generated around the well and connect with more natural fractures. Therefore, the directions of the maximum and minimum horizontal principal stresses should be considered in the design of horizontal well trajectories. Furthermore, the difference coefficient of the *in-situ* horizontal stress is an important factor affecting the morphology and complexity of fractures. In the case of a small difference coefficient, the extension direction of artificially induced fractures is consistent with that of natural fractures, making them easily connected to form complex fracture network systems. On the contrary, a larger difference coefficient of the *in-situ* stress will cause the direction of artificial fractures to gradually approach that of the maximum horizontal principal stress, making the connection between artificial and natural fractures harder, thus failing to form a fracture network system. In general, hydrocarbon migration and accumulation, wellbore instability while drilling, the design of horizontal well trajectories, reservoir stimulation, and well pattern deployment in the water injection stage are all related to *in-situ* stress. Therefore, accurate and effective prediction of the *in-situ* stress field of shale gas reservoirs is very significant for shale gas development with scale efficiency.

### 5.4.2. *Calculation of in-situ stress*

Foreign experts and scholars have carried out extensive work on the prediction and research of *in-situ* stress. Rock mechanical research can be traced back to the early 20th century. Firstly, the stress relief method was used to directly measured the rock stress in the tunnel under the Hoover Dam. Then, direct stress measurement methods such as the flat jack method, wellbore diametral deformation method, photoelastic stress meter method, and hydraulic fracturing measurement method were proposed successively until the 1960s. In the 1970s, Schlumberger began to attempt to use log data to solve geomechanical problems, including the determination of *in-situ* stresses, and applied logging methods to solve problems, such as layer collapse pressure, fracture pressure, and sand production in oil reservoirs. Price *et al.* (1990) deduced the quantitative relationship between stress, curvature, and elastic mechanical parameters, laying a theoretical foundation for estimating stress based on curvature. Moreover, they developed the curvature method for predicting *in-situ* stress, which is still widely used after modification and development by later scholars. Iverson (1995) first proposed an *in-situ* stress prediction model, which assumes that the horizontal direction of the strata is initially limited. In this case, there is no strain in the horizontal direction, and the *in-situ* stress field of the strata can be obtained by inversion based on the assumption of elastic strain. Gray *et al.* (2012) improved the empirical formula proposed by Iverson and proposed that the horizontal principal stress changes abruptly with changes in the elastic parameters of rock layers, and thus proposed a strain-based prediction method for the *in-situ* stress.

Chinese researchers also contributed to the prediction and research on *in-situ* stress. The research and application of *in-situ* stress measurement techniques began in China from the 1940s and the 1950s. Since the 1970s, China has successfully carried out *in-situ* stress measurements using the hydraulic fracturing method and improved deep-wellbore underwater three-directional strain gauges. It has successfully developed equipment and methods such as the piezomagnetic stress relief method and the Kaiser effect-based *in-situ* stress test. Relatively complete measurement and monitoring networks for *in-situ* stress have also been established in China. Huang *et al.* (1984) deduced a mathematical model for calculating the underground *in-situ* stress based on elastic mechanics under certain assumptions and boundary conditions (also referred to as the Huang Rongzun model) and used geophysical information to determine the model parameters. This method can be used for continuous calculation and analysis of *in-situ* stress and has been widely used in oil field production. Zhao *et al.* (1995) obtained the width and depth of wellbore breakouts using four-arm caliper curves and image log data and then calculated the maximum and minimum horizontal principal stresses according to rock mechanical properties. Around 2012, Fan and Deng *et al.* (2012) established an *in-situ* stress calculation method based on log data according to geomechanical principles, accurately deriving the *in-situ* stress around wells and their periphery. Zhang *et al.* (2015) proposed a process for establishing an equivalent petrophysical model of shale layers by comprehensively analyzing their anisotropic and fluid characteristics, and effectively predicted the minimum horizontal *in-situ* stress.

At present, there are four categories of methods for determining *in-situ* stress, namely, direct measurements, numerical simulations, calculations based on rock mechanics, and seismic predictions.

(1) **Methods of indirect estimations and direct measurements:** The fracture analysis method is a simple method for estimating the magnitude and direction of underground stress based on the concept that the formation and distribution of fractures are closely related to the *in-situ* stress. This method is based on Anderson's theory, in which the relative relationships among the minimum horizontal principal stress ($\sigma_h$), the maximum horizontal principal stress ($\sigma_H$), and the vertical stress ($\sigma_v$) are the main reasons for the formation of different types of fractures. If $\sigma_v > \sigma_H > \sigma_h$ or $\sigma_H > \sigma_h > \sigma_v$, a normal or a reverse fault is formed, respectively, the $\sigma_H$ is perpendicular to the fault strike, and $\sigma_h$ is parallel to the fault strike. If $\sigma_H > \sigma_v > \sigma_h$, a strike–slip fault is formed, the $\sigma_H$ is parallel to the fault strike, and the $\sigma_h$ is perpendicular to the fault strike. This method can be used to indirectly determine the stress state of rocks based on surface faults or faults predicted using other methods.

Hydraulic fracturing is a simple and adaptable method for the direct measurement of *in-situ* stress. The main steps of this method are as follows: first, select an interval to be measured in the target strata, isolate it with two suitably located rubber packers, then conduct hydraulic fracturing on this interval, and finally infer the stress state according to the extension of hydraulic fractures.

There are many methods for measuring *in-situ* stress, including direct measurements and indirect estimations. Besides the fault analysis method and the hydraulic fracturing method, the methods for direct measurements and indirect estimations also include the comprehensive geological mapping method, the sonic emission method, the core stress unloading method, the residual elastic strain recovery method, the stress recovery method, and the X-ray method.

(2) **Numerical simulation methods:** Based on large amounts of geological modeling as well as massive petrophysical, drilling, log, and engineering parameters, these methods simulate and predict the distribution and changes of the *in-situ* stress in underground media and determine the distribution patterns of the stress field through numerical simulations with the aid of high-performance computers. Depending on the theoretical basis and algorithms, the numerical simulation methods for *in-situ* stress prediction include the boundary load adjustment, boundary displacement analysis, stress function-based method, displacement inverse analysis, elastic mechanics, finite element method, and tectonic stress characteristic analysis.
(3) **Methods of log curve-based calculations:** The methods of indirect estimations and direct measurements can yield accurate *in-situ* stress. However, the small data volume, high costs, and discrete data make them impossible to produce continuous large-scale stress field data. Since the 20th century, experts and scholars led by petrophysicists have studied the methods for calculating the magnitude and direction of *in-situ* stress using log data to reduce costs and obtain continuous large-scale *in-situ* stress data. These methods are mainly based on sonic log, image log, and dipmeter log.

The method of sonic log-based calculation can be used to calculate both the magnitude and the direction of the *in-situ* stress based on shear-wave splitting. Specifically, a shear wave is split when the *in-situ* stress changes, and it propagates fast along the direction of the maximum. As a result, a fast shear wave is generated, and its azimuth is consistent with that of the maximum stress. This method mainly includes the following steps: (i) identify geological anomalies such as faults, fractures, and layer pinch-outs from image logs, and analyze possible causes of shear wave splitting, such as faults, fractures, bedding, and the related *in-situ* stress variations; (ii) process the multipole sonic log data to extract the shear wave information and determine the azimuth of the fast shear waves. (iii) determine the direction of the maximum principal stress after excluding non-*in-situ* stress factors from the azimuth of the fast shear waves. Rock mechanics parameters (e.g., Young's modulus and Poisson's ratio) can be calculated using compressional and shear waves velocities and density data. Afterward, the values of continuous *in-situ* stress can be obtained using the Huang Rongzun model.

With rock mechanics as the theorical basis, the method of image log-based calculation is mainly used to determine the magnitude and direction of the *in-situ* stress according to wellbore breakouts. According to rock mechanics, the tangential normal stress is maximum in the direction of the minimum principal stress. Accordingly, stress collapse or "wellbore breakout" is prone to occur in this direction, leading to the formation of an elliptical wellbore, whose long axis is in the same direction as the minimum horizontal principal stress. By contrast, the tangential normal stress is minimum in the direction of the maximum horizontal principal stress. When the density of drilling fluids is high during drilling, the rocks in the direction of the maximum principal stress will break due to tensile stress, resulting in induced fractures, whose strike is the direction of the maximum horizontal principal stress. Finally, the maximum and minimum horizontal principal stresses can be calculated according to rock mechanical properties using the width and depth of wellbore breakouts obtained using four-arm caliper curves and image logs, as follows:

$$\sigma_H = 2 \times \frac{(a_1 + a_2)(T - f\Delta p) - (c_1 + c_2)(T - e\Delta p)}{(a_1 + a_2)(d_1 + d_2) - (c_1 + c_2)(b_1 + b_2)} \tag{5.22}$$

$$\sigma_h = 2 \times \frac{(d_1 + d_2)(T - e\Delta p) - (b_1 + b_2)(T - e\Delta p)}{(a_1 + a_2)(d_1 + d_2) - (c_1 + c_2)(b_1 + b_2)} \tag{5.23}$$

where

$$a_1 = -\mu\left[1 - 2\cos\left(2\theta_\varphi\right) \times b\right]$$

$$a_2 = \left(1 + \mu^2\right)^{\frac{1}{2}}\left[1 - 2\cos\left(2\theta_\varphi\right) \times b\right]$$

$$b_1 = -\mu\left[1 + 2\cos\left(2\theta_\varphi\right) \times b\right]$$

$$b_2 = \left(1 + \mu^2\right)^{\frac{1}{2}}\left[1 + 2\cos\left(2\theta_\varphi\right) \times b\right]$$

$$c_1 = -\mu\left(1 + 2\frac{r_w^2}{D_{\max}^2}\right)$$

$$c_2 = \left(1 + \mu^2\right)^{\frac{1}{2}}\left[1 - \frac{r_w^2}{D_{\max}^2} + 3\frac{r_w^4}{D_{\max}^4}\right]$$

$$d_1 = -\mu\left(1 + 2\frac{r_w^2}{D_{\max}^2}\right)$$

$$d_2 = \left(1+\mu^2\right)^{\frac{1}{2}}\left(3\frac{r_w^2}{D_{\max}^2} - 3\frac{r_w^4}{D_{\max}^4} - 1\right)$$

$$e^2 = -\left(1+\mu^2\right)^{\frac{1}{2}}$$

$$f = -\left(1+\mu^2\right)^{\frac{1}{2}} \times \frac{r_w^2}{D_{\max}^2}$$

$$D_{\max} = |c_{13} - c_{24}|$$

where $\Delta p = p_m - p_p$, $p_m$ is the drilling fluid pressure, $p_p$ is the pore pressure, $T$ is the cohesion of rocks, $f$ is the internal friction coefficient of rocks, $c_{13}$ and $c_{24}$ are the wellbore minimum and maximum diameters, $\mu$ is the sliding friction coefficient of rocks, $D_{\max}$ is the maximum wellbore breakout depth, $b$ is the wellbore breakout width, $r_w$ is the nominal radius of the wellbore, and $\theta_\varphi$ is the azimuth of the intersection of the wellbore breakouts and the wellbore.

The maximum depth ($D_{\max}$) and width ($b$) of a wellbore breakout can be calculated using the dark strips on the image log, and the wellbore diameters ($c_{13}$ and $c_{24}$) can be obtained using the 1–3 caliper and 2–4 caliper curves recorded by the four-arm caliper tool.

Equations (5.22) and (5.23) were established based on the stable state of wellbore breakouts, and Liu *et al.* proposed the following equation to determine the stable state of a wellbore:

$$(\sqrt{1+f^2})\left[\sigma_h\left(1+2\frac{2r_1}{r_2}\right) - \sigma_h + 2(p_m - q)\frac{r_1}{r_2}\right] - 2fq < 2T \tag{5.24}$$

where $r_1$ and $r_2$ are the semi-minor axis and the semi-major axis of the elliptical wellbore, respectively; $q$ is the self-weight of rocks; and $T$ is the shear strength of rocks. Substituting the values of $D_{\max}$, $b$, and $r_w$ obtained from image logs and the calculated *in-situ* stress into Equation (5.24), the wellbore is stable if the equation is workable. In this case, the *in-situ* stress calculated based on wellbore breakouts is reliable.

The method of dipmeter log-based calculation has roughly the principle as the method of image log-based calculation, except for different logging tools and parameters to be measured. This method is mainly used to determine the wellbore breakout azimuth and the *in-situ* stress direction based on the caliper curves and the electrode azimuth curve in dipmeter logs. The main steps of this method are as follows: (i) determine the wellbore breakout interval according to the characteristics of the caliper curves and the electrode azimuth; (ii) infer the azimuth of the minimum horizontal principal stress according to the relative magnitude of the caliper curves and the electrode azimuth, and (iii) make statistics and determine the dominant azimuth of the minimum horizontal principal stress.

### 5.4.3. *Seismic-based in-situ stress prediction methods*

The abovementioned methods based on logs can be used to achieve the continuous calculation of *in-situ* stress and overcome the shortcomings of the methods of direct measurements such as scarce data points, failure in continuous measurement, and high costs. However, these methods can only obtain continuous calculation results from shallow to deep *in-situ* stress at well locations, that is, in close proximity to the well. With the exploration and development of shale gas and the demand for *in-situ* stress predictions, seismic-based *in-situ* stress prediction methods are increasingly applied and accepted. The major advantage of these methods is that they can obtain the continuous *in-situ* stress sections and data volumes of the 3D seismic coverage area and can accurately predict and describe the *in-situ* stress distribution in the 3D underground space. There are several methods for predicting the *in-situ* stress using seismic data.

(1) **Curvature-based prediction method:** Studies have shown that the change in *in-situ* stress is the main cause of the deformation, faulting, and fracturing of strata. Curvature directly reflects the degree of bending of strata and, thus, is an important indicator of fracture presence and *in-situ* stress anomalies. Therefore, curvature can be used to predict the distribution of *in-situ* stress. Many researchers the world over have studied the seismic prediction of *in-situ* stress. Price *et al.* deduced the quantitative relationship between *in-situ* stress, seismic curvature, and rock elastic parameters, laying the foundation for the prediction of the *in-situ* stress using curvature. Based on this, Sheorey *et al.*

researched the quantitative relationship between curvature, density, Young's modulus, and velocities of compressional and shear waves and proposed a new *in-situ* stress estimation model that considers the influence of the thermal expansion coefficient of rocks on the *in-situ* stress. Hunt (1990) derived a quantitative relationship between *in-situ* stress, curvature, and Young's modulus by analyzing the stratum factors affecting fracture density, providing a specific method for applying curvature to predict *in-situ* stress.

Chinese scholars have carried out extensive research and application work on curvature-based prediction of *in-situ* stress. He derived a calculation formula for the tectonic stress field based on the curvature analysis by simulating the tectonic stress field and established the tectonic stress field using curvature, providing a new idea for the simulation of the tectonic stress field and the prediction of the *in-situ* stress. Based on the prediction of fracture density and azimuth of buried-hill reservoirs, Tang and Li (2005) established the paleotectonic stress field and formed the seismic prediction method for the paleotectonic stress field using the finite element numerical simulation of continuous media, taking the rock failure criteria as constraints. Ma *et al.* (2018) applied the *in-situ* stress prediction method based on curvature and Young's modulus proposed by Hunt to shale gas reservoirs according to the following steps: (1) the curvature attributes of strata were extracted using the curvature attribute extraction method based on the Taup transform; (2) Young's modulus of strata was obtained through pre-stack seismic inversion, and the *in-situ* stress was determined using Hunt's formula. This application procedure provides a reference for the horizontal well deployment and the fracturing scheme design of shale gas development in the study area.

The basic principle of using curvature to predict *in-situ* stress is as follows: assuming that a bent rock layer has an original length of $L_0$, a top length of $L_1$, a bottom length of $L_2$, a thickness of $h$, and a radius of $R$, the top strain $e$ of the bent rock layer can be expressed as

$$e = \frac{L_1 - L_0}{L_0} \tag{5.25}$$

The curvature is defined as the reciprocal of the radius $R$, and the relational expression between strain and curvature can be expressed as

$$e = \frac{\frac{h}{2}}{R} = \frac{h}{2}K \tag{5.26}$$

where $K$ represents the maximum curvature.

The maximum positive curvature is the most effective curvature attribute in describing fractures, deflections, folds, and faults and can directly reflect the structural characteristics of the layer uplift. Therefore, the maximum positive curvature instead of the maximum curvature is selected to calculate the stress on strata. According to Hooke's law, the relationship between curvature and stress can be expressed as follows:

$$\sigma = Ee = E\left(\frac{h}{2}\right)K \tag{5.27}$$

Equation (5.27) can be further simplified as

$$\sigma = EK \tag{5.28}$$

According to Equation (5.28), the stress can be approximately equal to the product of Young's modulus and curvature, indicating that areas with large curvature and high Young's modulus have high stress, and are prone to fracturing and forming hydrocarbon accumulation spaces and migration channels. The *in-situ* stress prediction based on curvature can be conducted in three steps: (1) obtain the curvature attribute of reservoirs using the extraction method of seismic curvature attribute based on the local Taup transform; (2) extract Young's modulus data volume of the reservoirs through the pre-stack seismic inversion; (3) obtain the *in-situ* stress data volume by multiplying the curvature data volume by Young's modulus data volume according to Equation (5.28). This method can be used to quickly obtain the relative magnitude of the *in-situ* stress in a certain block. However, the direction of the *in-situ* stress can only be determined according to the regional stress direction or the trend surface of the *in-situ* stress magnitude and thus is not exactly consistent with the actual *in-situ* stress direction.

(2) **Horizontal strain model-based prediction method:** The horizontal strain model of strata used in this method was derived from the equation of layer strain at well locations. In its final expression, the maximum horizontal principal stress and the minimum horizontal principal stress are functions of parameters such as burial depth, Young's modulus, Poisson's ratio, and layer pressure:

$$S_h = \frac{v}{1-v}(S_v - \alpha p_p) + \frac{a}{1-v}S_v + \frac{A}{1-v^2}EH\varepsilon_h + \alpha p_p \quad (5.29)$$

$$S_H = \frac{v}{1-v}(S_v - \alpha p_p) + \frac{a}{1-v}S_v + \frac{B}{1-v^2}EH\varepsilon_h + \alpha p_p \quad (5.30)$$

where $S_h$ is the minimum horizontal principal stress, MPa; $S_H$ is the maximum horizontal principal stress, MPa; $S_v$ is the overburden pressure, MPa; $p_p$ is the pore pressure, MPa; $\alpha$ is the Biot coefficient; $v$ is Poisson's ratio; $a$ and $b$ are the regional stress coefficients in the directions of the minimum and maximum horizontal principal stresses, respectively; $A$ and $B$ are the contribution coefficients of the residual tectonic stress to the minimum and maximum horizontal principal stresses, respectively; $E$ is Young's modulus of rocks, GPa; $H$ is the layer depth, km; $\varepsilon_h$ is the dimensionless normalized strain coefficient of the minimum horizontal principal strain; and $\varepsilon_H$ is the dimensionless normalized strain coefficient of the maximum horizontal principal strain.

(3) **Anisotropic model-based prediction method:** The anisotropic model used in this study was derived from the strain model of strata taking into account the anisotropic factors. In its final expression, the maximum and minimum horizontal principal stresses are functions of parameters such as layer depth, Young's modulus, Poisson's ratio, layer pressure, and anisotropic normal flexibility:

$$S_h = \frac{v}{1-v}(S_v - \alpha p_p) + \frac{a}{1+EZ_n-v}(S_v - \alpha p_p) + \alpha p_p \quad (5.31)$$

$$S_H = \frac{v}{1-v}(S_v - \alpha p_p) + \frac{b(1+EZ_n)}{1+EZ_n-v}(S_v - \alpha p_p) + \alpha p_p \quad (5.32)$$

where $S_h$ is the minimum horizontal principal stress, MPa; $S_H$ is the maximum horizontal principal stress, MPa; $S_v$ is the overburden pressure, MPa; $p_p$ is the pore pressure, MPa; $\alpha$ is the Biot coefficient; $v$ is Poisson's ratio; $\alpha$ is the stress coefficient in the direction of the minimum horizontal principal stress; $b$ is the stress coefficient in the direction of the maximum horizontal principal stress; $E$ is Young's modulus of rocks, GPa; and $Z_n$ is the anisotropic normal flexibility, $\mathrm{GPa}^{-1}$.

Among these parameters, $Z_N$ is the normal flexibility in the linear sliding theory and can be obtained according to the relationship between the the normal flexibility and the normal weakness, as follows:

$$Z_N = \frac{\Delta_N}{(\lambda + 2\mu)(1 - \Delta_N)} \tag{5.33}$$

(4) **Combined spring model prediction method:** The model used in this study was developed based on the 3D elasticity theory and is the most commonly used model for horizontal stress estimation:

$$\sigma_h = \frac{\mu}{1-\mu}\sigma_v - \frac{\mu}{1-\mu}\alpha_{\text{vert}}p_p + \alpha_{hor}p_p + \frac{E}{1-\mu^2}\xi_h + \frac{\mu E}{1-\mu^2}\xi_h \tag{5.34}$$

$$\sigma_H = \frac{\mu}{1-\mu}\sigma_v - \frac{\mu}{1-\mu}\alpha_{\text{vert}}p_p + \alpha_{hor}p_p + \frac{E}{1-\mu^2}\xi_H + \frac{\mu E}{1-\mu^2}\xi_h \tag{5.35}$$

where $\sigma_H$ is the maximum horizontal stress; $\sigma_h$ is the minimum horizontal stress; $\sigma_v$ is the total vertical stress; $\alpha_{\text{vert}}$ is the effective stress coefficient in the vertical direction (Biot coefficient); $\alpha_{hor}$ is the effective stress coefficient in the horizontal direction (Biot coefficient); $\mu$ is the static Poisson ratio; $p_p$ is the pore pressure; $E$ is the static Young modulus; $\xi_h$ is the strain in the direction of the minimum principal stress; and $\xi_H$ is the strain in the direction of the maximum principal stress.

(5) **Huang Rongzun model-based on prediction method:** The Huang Rongzun model was derived and constructed based on the rock failure theory by Huang, a Chinese rock mechanical expert. In this model, the maximum and minimum horizontal principal

stresses are calculated using the fracturing data or fracturing experimental data, as follows:

$$\sigma_h = \left(\frac{v}{1-v} + A\right)(\sigma_z - \alpha p_p) + \alpha p_p \tag{5.36}$$

$$\sigma_H = \left(\frac{v}{1-v} + B\right)(\sigma_z - \alpha p_p) + \alpha p_p \tag{5.37}$$

where $A$ and $B$ are the dimensionless tectonic stress coefficients of rocks in the horizontal direction and can be obtained through the reverse calculation of hydraulic fracturing data; $\alpha$ is the dimensionless contribution coefficient of pore fluid pressure; and $p_p$ is the pore pressure.

(6) **Ge Hongkui model-based prediction method:** Professor Ge deduced the horizontal stress model and its expression that involve geomechanical parameters and tectonic stress parameters while comprehensively considering the influence of tectonic stress on the maximum and minimum horizontal principal stresses:

$$\sigma_h = \frac{v}{1-v}(\sigma_v - \alpha p_p) + K_h \frac{E(\sigma_v - \alpha p_p)}{1+v} + \frac{\alpha_T E \Delta T}{1-v} + \alpha p_p \tag{5.38}$$

$$\sigma_H = \frac{v}{1-v}(\sigma_v - \alpha p_p) + K_h \frac{E(\sigma_v - \alpha p_p)}{1+v} + \frac{\alpha_T E \Delta T}{1-v} + \alpha p_p \tag{5.39}$$

where $\alpha_T$ is the thermal expansion coefficient (zero if not considering temperature change); $\Delta_T$ is the change in layer temperature; $K_h$ and $K_H$ are the tectonic stress coefficients in the directions of the minimum and maximum horizontal formation stresses, respectively, and they can be regarded as constants within the same fault block.

(7) **Gray formula-based prediction method:** Gray *et al.* (2012) proposed a method for directly using wide-azimuth seismic data to estimate the principal stress of layers based on the constitutive equation of HTI media. This method can be used to obtain a continuous *in-situ* stress section of a certain area. This method can be used to calculate the *in-situ* stress in an area with few wells, as it does not need vast amounts of well data. Moreover, Gray

*et al.* (2012) proposed the concept of horizontal stress difference ratio (DHSR) to meet the needs of shale oil and gas development using horizontal wells. DHSR is an important parameter for evaluating whether a shale gas reservoir can be effectively fractured to create a fracture network. A smaller DHSR means that the reservoirs are more prone to be fractured to form a complex fracture network and yield higher production. This method can be used to obtain the continuous DHSR section of a certain area and then to find areas with low DHSR for hydraulic fracturing, thus optimizing the development of shale gas reservoirs. This is a new seismic-based method for obtaining *in-situ* stress data and can be expressed as follows:

$$\mathrm{DHSR} = \frac{EZ_N}{1 + EZ_N + v} \tag{5.40}$$

where DHSR is the horizontal stress difference ratio; $v$ is Poisson's ratio; $E$ is Young's modulus of rocks, GPa; and $Z_N$ is the anisotropic normal flexibility, $\mathrm{GPa}^{-1}$.

### 5.4.4. *Seismic prediction results of the in-situ stress*

Block Yang 101 in the study area, located in a low and steep tectonic zone in southern Sichuan, is important for shale gas development. As shown in Figure 5.14, this block appears as an NE-trending band and is characterized by alternating horsts and grabens. A number of low and steep anticlines are present and subdivide the block, accordingly. As a result, this block consists of four uplifts and four depressions from east to west. The uplifts have an axial direction of nearly north-south, are narrow and steep, and consist of a series of faulted anticlines. The negative structures consist of multiple wide and gentle synclines.

There are two sets of faults in the study area, with one set having a nearly NS strike and the other having an NE-SW strike. The NS-trending faults have a larger scale and predominate in the study area, and the NE-SW trending faults were formed earlier.

All the faults in the study area are reverse faults. As preliminarily determined based on the relationship between faults and stresses, the

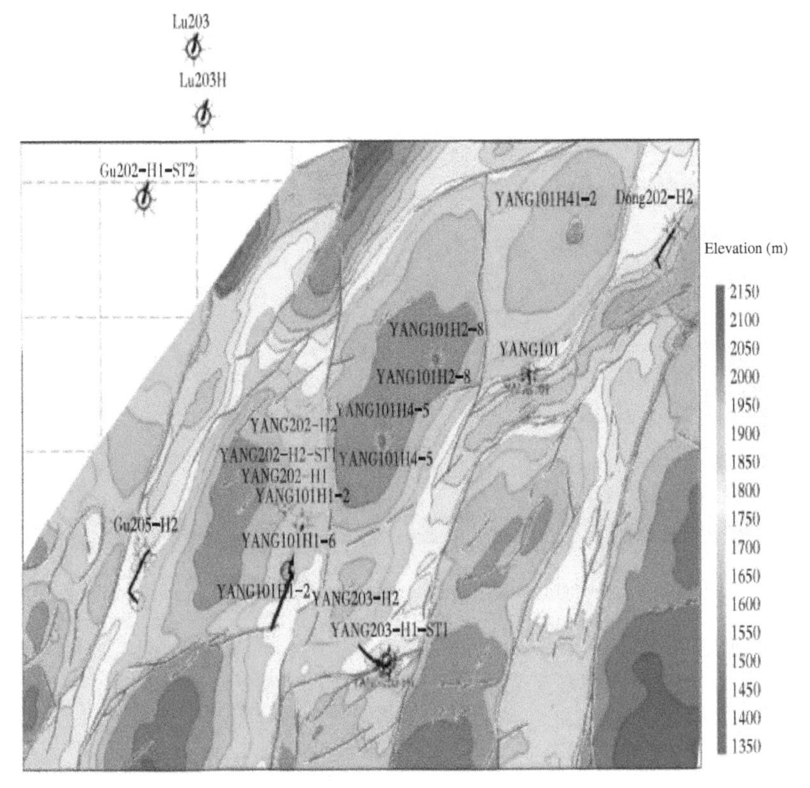


**Figure 5.14.** Structural map of the bottom of the Longmaxi Formation in block Yang 101.

maximum and minimum principal stresses of the study area trend nearly EW and nearly NS, respectively.

This study adopted a poroelastic horizontal strain model to calculate the maximum and minimum horizontal stresses and the horizontal stress difference of this study area using vertical stress, pore stress, and Poisson's ratio based on the 3D elastic theory. Figures 5.15–5.17 show the maps of the maximum horizontal principal stress, minimum horizontal principal stress, and horizontal stress difference of the target intervals in the study area, respectively.

The predicted results of the Long-$I_1$ Submember in block Yang 101 are as follows: (1) the maximum principal stress in this submember is concentrated in the wide and gentle structural area

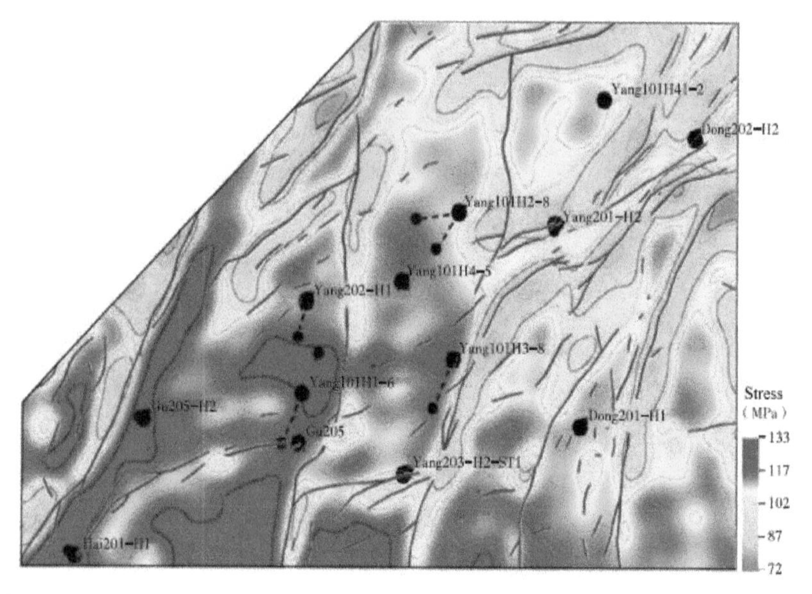


**Figure 5.15.** Map of the maximum horizontal principal stress of the Long-$I_1$ Submember in block Yang 101.

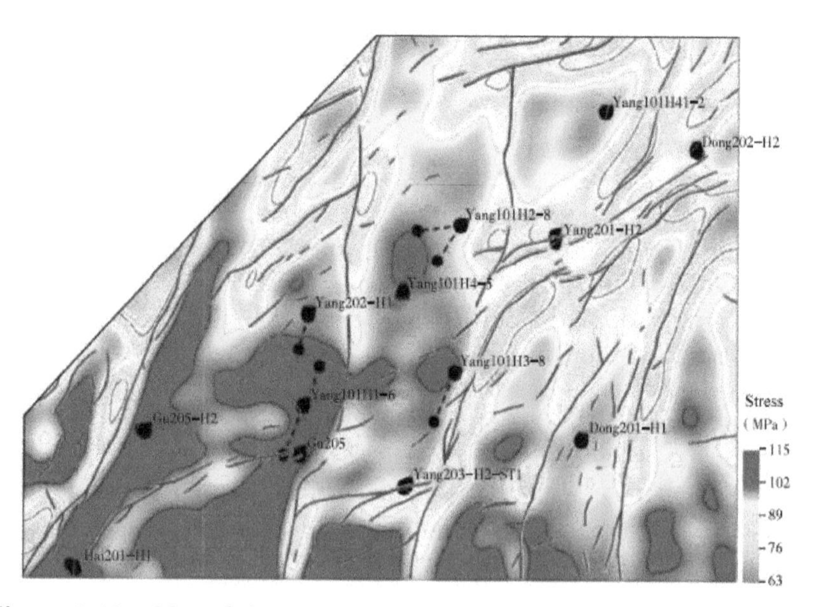


**Figure 5.16.** Map of the minimum horizontal principal stress of the Long-$I_1$ Submember in block Yang 101.

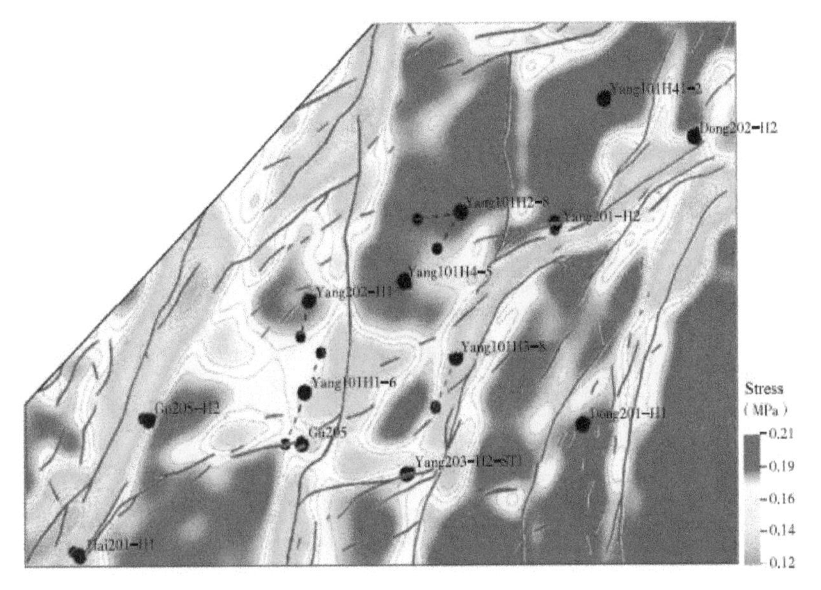


**Figure 5.17.** Map of the horizontal stress difference of the Long-$I_1$ Submember in block Yang 101.

under the influence of structures and fractures, with a maximum value of approximately about 72–133 MPa. By contrast, fractures are developed and pressure is released at higher structural positions (faulted anticlines) under the action of tensile stress; (2) the minimum principal stress has an overall trend similar to the maximum horizontal stress. Specifically, it is released at higher structural positions, with a minimum value of approximately 63 MPa, and concentrated in wide and gentle structural areas, with a maximum of up to 115 MPa; (3) Owing to the influence of the narrow and steep faulted anticlines, the wide and gentle structural zones have a smaller horizontal stress difference, with a stress difference coefficient under 14%. Therefore, they are favorable for effective hydraulic fracturing. In addition, local areas show a relatively high stress difference due to the influence of faults.

It is also essential to determine the direction of the *in-situ* stress based on the horizontal strain model of a strata. For a stressed hexahedron in the geological body, its internal points will be subject to displacement along the coordinate axes under stress. Assuming that the displacements along the $X$, $Y$, and $Z$ coordinate axes are $\mu$, $v$, and $\omega$, respectively, they are functions of $x$, $y$, and $z$ as follows:

$$\begin{cases} \mu = u(x,y,z) \\ v = v(x,y,z) \\ w = w(x,y,z) \end{cases}$$

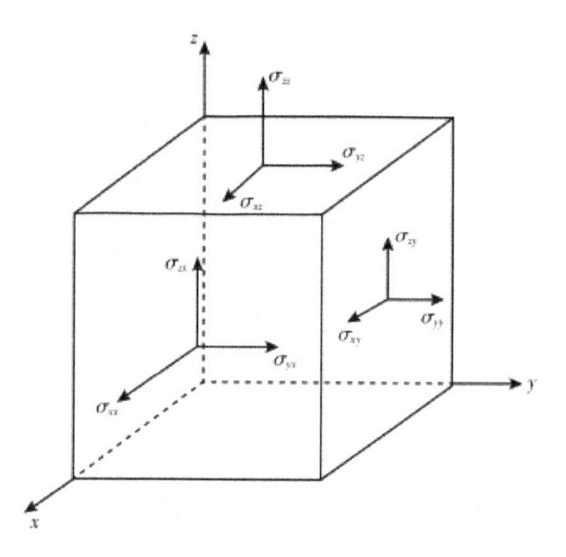


It is assumed that the strains along the three axes are $\varepsilon_x$, $\varepsilon_y$, and $\varepsilon_z$, respectively, the shear strains on the three planes are $\gamma_{xy}$, $\gamma_{yz}$, and $\gamma_{zx}$, respectively, the triaxial normal stresses are $\sigma_x$, $\sigma_x$, and $\sigma_z$, respectively, and the triaxial shear stresses are $T_{xy}$, $T_{yz}$, and $T_{zx}$, respectively.

According to the theory of elastic mechanics, the relationships between the strains and displacements above are as follows:

$$\begin{cases} \varepsilon_x = \dfrac{\partial \mu}{\partial x}, \quad \gamma_{xy} = \gamma_{yx} = \dfrac{\partial u}{\partial y} + \dfrac{\partial v}{\partial x} \\ \varepsilon_y = \dfrac{\partial v}{\partial y}, \quad \gamma_{yz} = \gamma_{zy} = \dfrac{\partial v}{\partial z} + \dfrac{\partial w}{\partial y} \\ \varepsilon_z = \dfrac{\partial w}{\partial z}, \quad \gamma_{zx} = \gamma_{xz} = \dfrac{\partial w}{\partial x} + \dfrac{\partial u}{\partial z} \end{cases} \tag{5.41}$$

The relationships between stresses and strains are as follows:

$$\begin{cases} \varepsilon_x = \dfrac{1}{E}\left[\sigma_x - \mu\left(\sigma_y + \sigma_z\right)\right], & \gamma_{xy} = \dfrac{T_{xy}}{G} \\ \varepsilon_y = \dfrac{1}{E}\left[\sigma_y - \mu\left(\sigma_x + \sigma_z\right)\right], & \gamma_{yz} = \dfrac{T_{yz}}{G} \\ \varepsilon_z = \dfrac{1}{E}\left[\sigma_z - \mu\left(\sigma_x + \sigma_y\right)\right], & \gamma_{zx} = \dfrac{T_{zx}}{G} \end{cases} \tag{5.42}$$

Their inverse relationships are as follows:

$$\begin{aligned} \sigma_x &= 2G\varepsilon_x + \lambda\theta, \quad T_{xy} = G\gamma_{xy} \\ \sigma_y &= 2G\varepsilon_y + \lambda\theta, \quad T_{yz} = G\gamma_{yz} \\ \sigma_z &= 2G\varepsilon_z + \lambda\theta, \quad T_{xz} = G\gamma_{xz} \\ \theta &= \varepsilon_{kk} \end{aligned}$$

where $\lambda$ is the Lamé constants, $G$ is the shear modulus, $E$ is Young's modulus, $\theta$ is the volumetric strain, and $\mu$ is Poisson's ratio.

Assuming that:

(1) The thickness ($h$) of the rock layer is much smaller than its lateral dimensions;
(2) The rock layer is in the horizontal state before deformation, and its state during the deformation is not considered;
(3) The deformation of the layer is caused by lateral tensile stress, and vertical stress is not considered;
(4) Rocks are isotropic;
(5) Rocks are in the linear elastic deformation range;
(6) Relaxation is not considered, the triaxial stress field can be simplified to the 2D stress field according to the thin plate theory, hence,

$$u_x = z\frac{\partial w}{\partial x}, \quad u_y = z\frac{\partial w}{\partial y} \tag{5.43}$$

and

$$\varepsilon_x = z\frac{\partial^2 w}{\partial x^2}, \quad \varepsilon_y = z\frac{\partial^2 w}{\partial y^2}, \quad \gamma_{xy} = 2z\frac{\partial^2 w}{\partial x \partial y}$$

The curvature deformation components are defined as

$$K_x = -\frac{\partial^2 w}{\partial x^2}, \quad K_y = -\frac{\partial^2 w}{\partial y^2}, \quad K_{xy} = -\frac{\partial^2 w}{\partial x \partial y} \tag{5.44}$$

Thus, the strain components can be expressed as

$$\varepsilon_x = -zk_x, \quad \varepsilon_y = -zk_y, \quad \gamma_{xy} = -2zk_{xy} \tag{5.45}$$

Substituting the previous equations yields the following:

$$\sigma_x = -\frac{Eh}{2(1-\mu^2)}(k_x + \mu k_y), \quad \sigma_y = -\frac{Eh}{2(1-\mu^2)}(k_y + \mu k_x),$$

$$T_{xy} = -\frac{Eh}{2(1+\mu)}k_{xy} \tag{5.46}$$

Substituting the layer thickness $h = 2z$ into Equation (5.46) yields the stress components on the layer plane represented by the curvature components as follows:

$$\mu_z = \frac{\varepsilon}{1-\mu^2}(\varepsilon_x + \mu\varepsilon_y), \quad \sigma_y = \frac{E}{1-\mu^2}(\varepsilon_x + \mu\varepsilon_x), \quad T_{xy} = Gy_{xy} \tag{5.47}$$

According to Equation (5.47), when the layer plane is convex upward, the curvature is greater than 0, corresponding to the positive tensile stress on the layer plane that curves outward.

After the stress of a point along the three coordinates is obtained, the principal stress and its direction can be calculated as follows:

$$\sigma_{\max} = \frac{\sigma_x + \sigma_y}{2} + \sqrt{\left(\frac{\sigma_x - \sigma_y}{2}\right)^2 + T_{xy}^2},$$

$$\sigma_{\min} = \frac{\sigma_x + \sigma_y}{2} - \sqrt{\left(\frac{\sigma_x - \sigma_y}{2}\right)^2 + T_{xy}^2} \tag{5.48}$$

Angle $\alpha$ between $\sigma_{\max}$ and $X$ axis and angle $\beta$ between $\sigma_{\min}$ and $X$ axis are as follows:

$$t_a = \frac{\sigma_{\max} - \sigma_x}{T_{xy}}, \quad t_g(\beta) = \frac{\tau_{xy}}{\sigma_{\min} - \sigma_y} \tag{5.49}$$

According to Equation (5.49), the stress field at the point can be estimated using the disturbance equation of the layer plane or

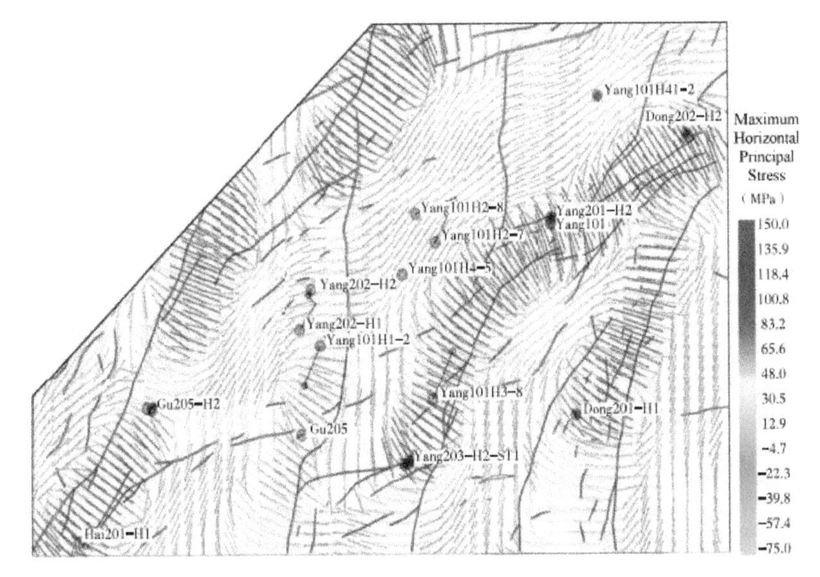


**Figure 5.18.** Map of the direction of the maximum horizontal principal stress of the Long-$I_1$ Submember in block Yang 101.

the curvature of the point. Generally, the direction of the maximum horizontal stress is defined as the stress direction in the 2D stress field.

The maximum principal stress of the Long-$I_1$ Submember in block Yang 101 is affected by structures and faults. At the high structural positions (faulted anticlines), faults are developed under the action of tensile stress, and the maximum principal stress is perpendicular to the strike of faults and in the NW-SE direction. By contrast, the gentle structural zones are affected by stresses from multiple directions and are dominated by compressive stress (Figure 5.18).

The minimum principal stress of the Long-$I_1$ Submember in block Yang 101 is in the NE-SW direction overall. It is parallel to the strike of the faults at higher structural positions due to the action of tensile stress and turns to the NW-SE direction or nearly W-E direction in the gentle structural zones dominated by compressive stress (Figure 5.19).

Figure 5.20 shows the superposition map of the difference coefficient of the stress, the maximum horizontal principal stress direction, and the fault prediction results for the bottom layer of the Longmaxi

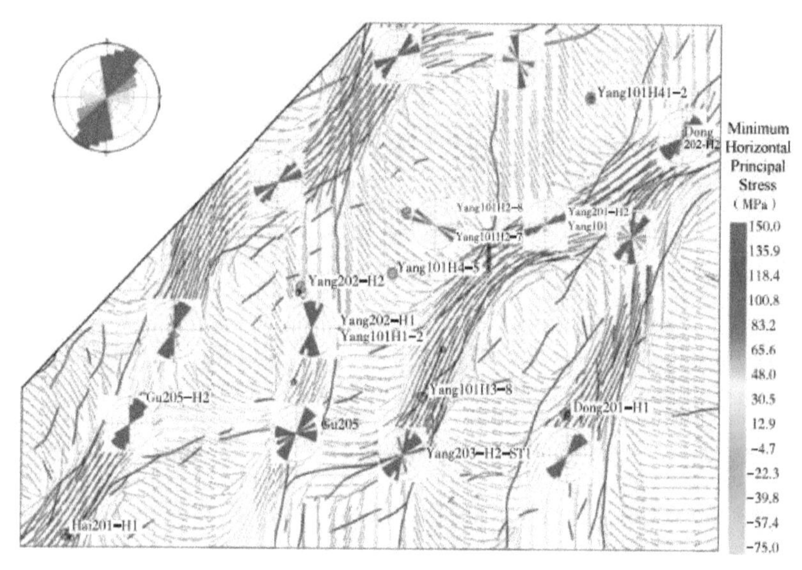


**Figure 5.19.** Map of the direction of the minimum horizontal principal stress of the Long-$I_1$ Submember in block Yang 101.

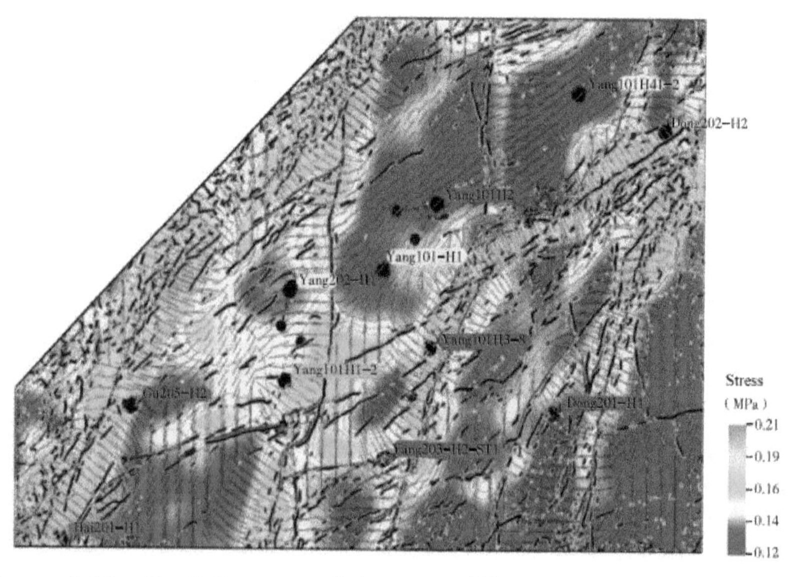


**Figure 5.20.** Prediction map of the stress difference superimposed with the maximum horizontal principal stress direction for the Long-$I_1$ Submember in block Yang 101.

Formation in block Yang 101. The horizontal stress difference is relatively consistent in the study area, with a difference coefficient of 0.12–0.21 and less than 0.14 for most of the area. Overall, the wide and gentle structural zones have small horizontal stress difference and are subject to stresses from multiple directions. Moreover, the maximum principal stress direction is perpendicular to the faults' strike and in the NW-SE direction, and there are regions with relatively high stress difference locally.

The seismic prediction of *in-situ* stress, which requires large amounts of input data and advanced techniques, is a new field of seismic interpretation. At present, there are only a few pieces of commercial software that can be directly used for the seismic prediction of the *in-situ* stress and they are yet to be improved. Therefore, there may be a certain gap between the accuracy of seismic prediction of *in-situ* stress and the required accuracy, and further improvements are needed.

## 5.5. Summary

(1) Rock mechanical parameters such as brittleness play an important role in controlling the productivity of shale gas reservoirs, compared to conventional reservoirs. Among the numerous characterization parameters and methods of rock brittleness, the brittleness index can reflect the actual compressibility of strata. Moreover, the brittleness index is derived from large amounts of test and measurement data according to the principle of rock mechanics in shale gas development and is suitable for the brittleness description of shale reservoirs.

(2) As indicated by the seismic prediction results of the brittleness index, the brittleness index of rocks in the Changning block is greater than 50% generally and is consistent laterally, making this block the most favorable area in terms of fracturing quality and suitable for later re-fracturing. The regions with high brittleness index are mainly distributed in the syncline cores in the central portion of the study area, supporting the view that the brittle minerals of marine shale gas reservoirs in this area are mainly endogenetic deposits.

(3) The density of natural fractures greatly influences the performance of reservoir fracturing stimulation. The curvature attribute can not only reflect the anisotropy of reservoirs but also embodies the tectonic stress of the rock layers. The curvature-based fracture prediction technique has high prediction precision and is compatible with the seismic data of this area, thus featuring both accuracy and applicability. Therefore, this technique is applicable for fracture prediction in this area.
(4) The fracture prediction results show that the natural fractures are widely and consistently developed in the study area. The widely developed natural fractures in the Changning block have a positive influence on the stimulation performance of the shale gas reservoirs in this block and are an important factor contributing to the relatively superior shale gas reservoirs in this area. Moreover, there are two fracture sets with highly consistent orientations, namely, a dominant set oriented nearly NW-SE and a secondary set oriented nearly NE-SW.

## Bibliography

Deng JG, Zhu HY, Xie YH, *et al.* Study on the characteristics and fracturing mechanisms of difficult to drill strata in the Western South China Sea. *Geotechnical Mechanics*, 2012, 33 (07): 2097–2102+2109.

Gray D, Anderson P, Logel J, *et al.* Estimation of stress and geomechanical properties using 3D seismic data. *First Break*, 2012, 30 (3): 59–68.

He Y, He ZH, Xiong XJ. Fault identification method based on high-precision curvature analysis. *Journal of Petroleum and Natural Gas*, 2010, 32 (06): 404–407+541.

Higgins S, Goodwin S, Donald A, *et al.* Anisotropic stress models improve completion design in the Baxter shale. *SPE115736*, 2008: 321–329.

Huang RZ. A model for predicting formation fracture pressure. *Journal of the University of Petroleum*, 1984 (4): 335–347.

Hunt JM. Generation and migration of petroleum from abnormally pressured fluid compartments. *GeoScienceWorld*, 1990, 74 (1): 1–12.

Iverson WP. Closure stress calculations in anisotropic formations. Paper presented at the Low Permeability Reservoirs Symposium, Denver, Colorado, March 1995.

Price NJ and Cosgrove JW. *Analysis of Geological Structures.* Cambridge University Press, Cambridge, 1990.

Rickman R, Mullen M, Petre J, *et al.* A practical use of shale petrophysics for stimulation design optimization: All shale plays are not clones of the Barnett Shale. In *SPE Annual Technical Conference and Exhibition*, 21–24, Denver, Colorado, USA, September 2008.
Rivera SR, Hhandwerger D, Kieschnick J, *et al.* Accounting for heterogeneity provides a new perspective for completions in tight gas shales. Paper presented at the *Alaska Rocks 2005, The 40th U.S. Symposium on Rock Mechanics (USRMS)*, Anchorage, Alaska, June 2005.
Safdar K, Sajjad A, Han HX, *et al.* Importance of shale anisotropy in estimating in-situ stresses and wellbore stability analysis in Horn river basin. In *Paper presented at the Canadian Unconventional Resources Conference*, Calgary, Alberta, Canada, November 15–17, 2011.
Sheorey PR. A theory for insitu stresses in isotropic and transverseley isotropic rock. *International Journal of Rock Mechanics and Mining Sciences*, 1994, 31 (1): 23–34.
Tang XR, Li J. Application of finite element numerical simulation of tectonic stress field in fracture prediction. *Special Oil and Gas Reservoirs*, 2005, 12 (2): 25–28.
Zhao LX. Analyzing the mechanism and characteristics of fracturing leakage and wellbore stress collapse using logging data. *Drilling and Completion Fluids*, 1995, (01): 17–24.
Zhang GZ, Chen JJ, Chen HZ, *et al.* Prediction for *in-situ* formation stress of shale based on rock physics equivalent model. *Chinese Journal of Geophysics*, 2015, 58 (6): 2112–2122.

https://doi.org/10.1142/9789811283185_0006

# Chapter 6

# Comprehensive Quantitative Prediction of Shale Gas Sweet Spots

The distribution of shale gas sweet spots is affected and controlled by multiple and interconnected factors. Understandably, the multi-map superposition method used for the comprehensive evaluation of conventional oil and gas reservoirs is not suitable for an accurate prediction and evaluation of shale gas sweet spots. This chapter first summarizes and compares the methods for the comprehensive prediction of shale gas sweet spots used at home and abroad and analyzes their advantages and disadvantages. Then, based on the fuzzy optimization principle, this study establishes a model for the quantitative prediction of shale gas sweet spots in southern Sichuan and then comprehensively evaluates shale gas sweet spots in the Changning block and predicts their distribution using this model.

## 6.1. Methods for comprehensive quantitative prediction of reservoirs

### 6.1.1. *Overview*

In recent years, with the commercial development of shale gas in North America, thanks to many breakthroughs in core technologies, such as horizontal wells, sweet spots prediction, hydraulic fracturing, and microseismic monitoring as well as the continuous improvements of related supporting technologies, shale gas has become one of the main foci of hydrocarbon exploration and development around the world. China enjoys enormous shale gas potential and large prospects for shale gas exploration and development. Ranking and selecting shale gas sweet spots from large numbers of exploration targets, as well as determining their distribution and quantitative characterization, have become hot and complex problems in research on shale gas exploration and development techniques. The already available technologies and methods to characterize, evaluate, and rank shale gas sweet spots developed abroad cannot be readily applied in China due to the complex geological history characterized by multiple tectonic phases of this country. Tailored sweet spots prediction methods, especially the seismic prediction methods, need to be developed in China to determine the sweet spots for the development with scale efficiency and meet the demand for natural gas development in China.

This chapter first summarizes and compares comprehensive prediction methods for shale gas sweet spots in China and globally. Then, it analyzes the main factors controlling shale gas sweet spots and the technical factors affecting their prediction while considering the problems existing in China's shale gas exploration and development practice. Based on this, this study proposes a new and alternative method for quantitative prediction of shale gas sweet spots, which has successfully been applied in the study area.

### 6.1.2. *Comparison of methods*

Comprehensive quantitative prediction and evaluation methods are critical for both conventional and unconventional oil and gas exploration. Different companies have adopted different prediction methods and quantitative prediction parameter systems for shale gas sweet spots based on the specific geological conditions of oil and gas zones as well as the stages and techniques of shale gas exploration and development. Three methods are commonly used at present, namely, the comprehensive risk analysis method, the boundary network node method, and the geological parameter-based multi-map superposition method. The first method was developed by Newfield Exploration Company and BP, the second method is supported by ExxonMobil, and the third method has been widely applied by many energy companies, including Harding Shelton Group, HESS, and Chevron. Even for the same method, different companies use different technologies and parameter systems in different areas, with the parameter systems consisting of different parameters and parameter values (Table 6.1).

A comprehensive analysis of these methods reveals that the comprehensive risk analysis method is generally suitable for mature exploration areas with complete data. When predicting and evaluating shale gas sweet spots using the comprehensive risk analysis method, BP Company mainly considers nine reservoir parameters, including not only geological factors such as the thermal maturity of organic matter, TOC content, thickness, distribution area, and matrix porosity of shale gas reservoirs but also engineering factors such as formation pressure and brittle mineral content.

ExxonMobil comprehensively predicts the distribution of shale gas sweet spots using 13 parameters according to the node network analysis method, which attaches great importance to the economic benefits in the application and takes the economic limit production of shale gas wells as the objective function in the prediction of shale gas sweet spots. Moreover, the 13 parameters are analyzed as boundary functions that affect the economic parameters in this method. Since this method mainly considers the influence of geological and engineering factors on economic benefits, it cannot objectively reflect the actual exploration and development potential of shale gas reservoirs in many cases.

Harding Shelton Energy Company considers up to 16 parameters in the prediction and comprehensive evaluation of shale gas sweet spots. These parameters roughly include three categories, namely, geological factors (e.g., the structure, TOC content, and thickness of reservoirs), environmental factors (e.g., water sources and water treatment), and drilling factors (e.g., well site conditions and natural gas pipeline networks).

In China, many researchers have done massive research on the comprehensive prediction and evaluation methods of shale gas. For instance, the Resources Strategic Research Center of Oil & Gas under

**Table 6.1.** Shale gas evaluation parameters used by some oil companies in China and globally.

| No. | Company | Evaluation indicators | Number |
|---|---|---|---|
| 1 | Harding Shelton Group | Geological factors: the total organic carbon (TOC) content, cumulative thickness, brittle mineral content, porosity, continuity, permeability, pore pressure, and pressure gradient of shale gas reservoirs; mineral composition; the thermal maturity of organic matter; the quality of raw seismic data; and structural complexity.<br>Drilling factors: well sites, the surface conditions at well sites, pipe network conditions.<br>Environmental factors: distance from water sources, water treatment conditions, and environmental assessment and its influencing factors. | 16 |
| 2 | ExxonMobil | The TOC content, burial depth, lithology, brittle mineral content, brittle mineral compressibility, and area of reservoirs; average TOC content of strata; the thermal maturity of organic matter; the density and types of fractures; the effective thickness of high-quality shale reservoirs; the types and sizes of matrix pores; the content and distribution of other non-hydrocarbon gases; and proportions of adsorbed gas and free gas. | 12 |
| 3 | BP | Paleostructural features and basin evolution history; the sedimentary facies and subfacies, thickness, TOC content, vitrinite reflectance ($Ro$), brittle mineral content, burial depth, and structures of shale gas reservoirs; geothermal gradient, and geothermal temperature. | 9 |
| 4 | Chevron | TOC content; the thermal maturity of organic matter; the effective thickness, rock brittleness, burial depth, and pore pressure of shale gas reservoirs; original sedimentary environment; and structural complexity. | 8 |
| 5 | SINOPEC | TOC content, the thickness of high-quality shale gas reservoirs, brittle mineral content, formation pressure, paleosedimentary environment, and the scale and distribution of faults. | 6 |

the Ministry of Land and Resources determined three evaluation classes of shale gas blocks. The first class is prospecting areas, such as the upper Yangtze area in south China; the second class is favorable areas, such as the southern Sichuan in south China, and the third class is core or target areas, such as the Changning, Weiyuan, and Fuling areas. The Center also divides the shale gas in China into three types according to lithology, namely, continental, marine, and marine-continental transitional shale gas. Different parameter systems will be used for predicting and ranking the comprehensive prediction and type-based evaluation methods of shale gas.

In its shale gas development practice in southern Sichuan, SINOPEC has determined six important factors for shale gas sweet spot prediction, including the TOC content and brittle mineral content of shale reservoirs, the continuous thickness of high-quality reservoirs, the pore pressure of strata and its preservation conditions, sedimentary environment and sedimentary facies zones, and the scale and distribution of faults. The rough weight of each of these factors in the final evaluation has been preliminarily determined according to the experience of experts, and the final comprehensive evaluation results were obtained by weighted superposition of the calculation results of individual factors.

### 6.1.3. *Quantitative prediction and ranking of shale gas sweet spots*

As shown by the shale gas exploration and development experience in the United States and Canada, the shale gas development with scale efficiency is jointly influenced by geological conditions and the development and utilization conditions. Given the actual situation of shale gas exploration and development in China and the characteristics and distribution of its major shale gas reservoirs, the abovementioned two factors also play a major role in the shale gas development with scale efficiency in China. Since China has just started shale gas development and all conditions for development and utilization are at the initial stage, it is more significant to consider the geological conditions. The accurate identification of sweet spots is the most important for the achievement of industrial productivity. Therefore, parameters related to geological conditions are primarily selected for the prediction and evaluation of shale gas sweet spots, followed by parameters related to the development and utilization conditions. Among them, the selected geological conditions-related parameters include the continuous thickness of effective reservoirs, TOC content, the thermal maturity of organic matter, gas content, and porosity, while the selected parameters related to the development and utilization conditions include structural conditions and the burial depth and hydrological conditions of reservoirs. Table 6.2. shows the specific evaluation indicators, their values, and the effects of different values on the evaluation results.

Many parameters in Table 6.2 cannot be quantified, such as tectonic conditions, hydrological conditions, and market demand. They tend to originate from many sources and are expressed in different units. Moreover, the classes (value ranges) of each parameter are mostly determined according to the experience of experts rather than on a theoretical basis. All these make the final evaluation results difficult to quantify and lead to low accuracy. Therefore, this parameter system is only suitable for the characterization and ranking of shale gas exploration areas.

According to the successful experience in shale gas exploration and development in China and globally, the geological factors controlling shale gas accumulation and sweet spot enrichment mainly include the burial depth, TOC content, physical properties and thickness of shale gas reservoirs, the maturity of organic matter, gas content, formation pressure, *in-situ* stress, rock mechanical properties, lithological associations, shale gas preservation conditions, the development degree of natural fractures, and structural complexity. Among them, favorable parameter values include TOC content >2%, $R_o$ of 1.43.0, burial depth <4,000 m, effective thickness >30 m, and brittle mineral content >40%. Figure 6.1 shows various parameters and their ideal value ranges.

Although Figure 6.1 gives the ideal value ranges of various geological factors that affect sweet spots, these values are used to determine favorable areas in actual production. Owing to the strong heterogeneity of shale gas reservoirs in China, it is difficult to select the optimal sweet spots for the development with scale efficiency from the favorable areas. This is primarily caused by the fact that these evaluation parameters are difficult to quantify.

**Table 6.2.** Domestic parameter system shale gas evaluation.

| Evaluation conditions | Indicator | Class of indicator (value range) | | | |
|---|---|---|---|---|---|
| | | **100–75** | **75–50** | **50–25** | **25–0** |
| Geological conditions | Abundance of resources (108 $m^3/km^2$) | >1.5 | 1.5–1.0 | 1.0–0.5 | <0.5 |
| | Effective thickness (m) | >75 | 75–50 | 50–30 | 30–10 |
| | Abundance of organic matter (%) | >2 | 1.0–2.0 | 0.5–1.0 | 0.3–0.5 |
| | Maturity (%) | >2 | 1.2–2.0 | 0.5–1.2 | <0.5 |
| | Permeability (mD) | >0.5 | 0.2–0.5 | 0.1–0.2 | <0.1 |
| | *In-situ* stress (MPa) | >22 | 22–15 | 15–10 | <10 |
| | Quantity of adsorption (%) | >85 | 60–85 | 40–60 | <40 |
| | Development degree of fracture | Well developed | Developed | Poorly developed | Extremely poorly developed |
| Development conditions | Tectonic conditions | Very favorable | Favorable | Moderate | Poor |
| | Hydrological conditions | Very favorable | Favorable | Moderate | Poor |
| | Burial depth (m) | >2,000 | 1,500–2,000 | 500–1,500 | <500 |
| | Terrain conditions | Plain | Hill, plateau | Tableland, slope | Desert, gobi |
| | Market demand | Very high | High | Moderate | Low |
| | Infrastructure | Perfect | Good | Moderate | Poor |

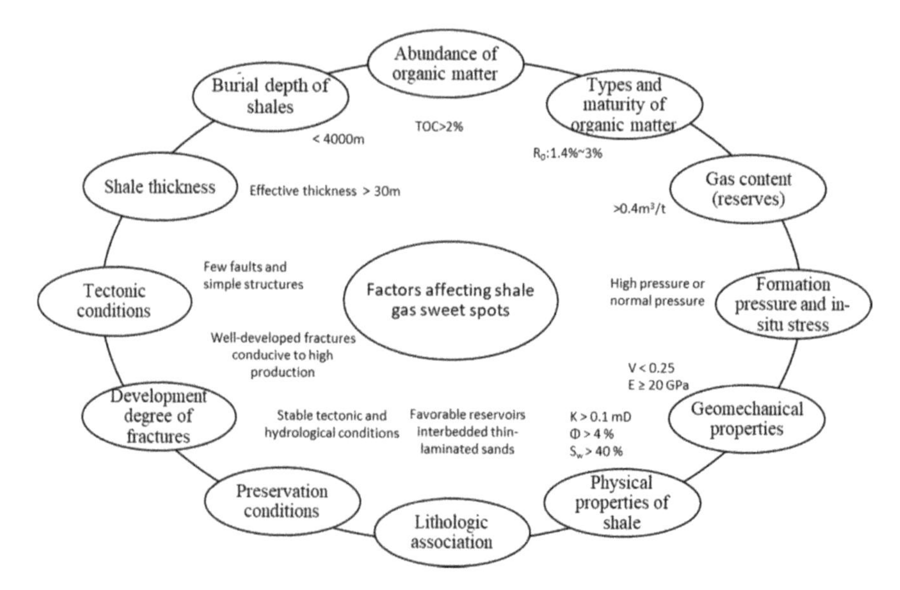


**Figure 6.1.** Factors affecting shale gas sweet spots.

## 6.2. A new method for seismic quantitative prediction of shale gas sweet spots

China has only made breakthroughs in shale gas development in some blocks in southern Sichuan and is still in the initial stage of industrialized shale gas exploration and development overall. Specifically, no breakthrough has been made in core technologies so far, the basic theoretical research is tenuous, and key technologies such as comprehensive quantitative evaluation are just being developed. Many domestic research institutions and colleges have predicted and evaluated China's shale gas resources or reserves and identified some favorable areas. However, a true shale gas sweet spot prediction and evaluation system for sweet spot exploitation and horizontal well deployment has not been established yet.

The prediction of core shale gas development areas should consider both the geological conditions of reservoirs and engineering factors. Multiple controlling factors have complex relationships with each other and are interconnected. Moreover, they originate from various sources and their weights related to the effects on sweet spots are difficult to quantify. Therefore, this study developed an equation for the quantitative comprehensive sweet spot evaluation by introducing the fuzzy optimization theory and fully utilizing geological, engineering, and production data. Then, with the individual factors as inputs and production data as output, this study automatically optimized the equation by adjusting the optimization conditions according to actual production conditions and determined the weights of various individual parameters in the prediction results and their upper and lower limits. In this manner, a quantitative sweet spot prediction system was built.

### 6.2.1. *Principle of fuzzy optimization theory*

With the advances in mathematics and computing techniques, the solutions for many systems with clear definitions and structures have been rapidly developed and gradually matured. However, there are still many problems that cannot be quantitatively described using theories and methods due to the heterogeneity of data sources and interdependencies between data (unclear boundaries). The comprehensive prediction and evaluation of shale gas sweet spots are one such problem due to its characteristics of numerous primary controlling factors, multiple methods for solving a single factor, diverse data sources, the complex relationships and interdependencies between the factors, hardly definable degree of control of sweet spots by various factors, and unclear boundaries.

Fuzzy optimization provides an effective means for this kind of problem. Based on fuzzy mathematics, fuzzy optimization defines the optimal evaluation criteria according to shale gas production demand and then prioritizes all parameters and gives the quantitative results, which are used as the basis for analysis and evaluation.

In the fuzzy optimization method, fuzzy subsets are first built through the analysis of reservoir characteristics and production conditions to quantify the factors or evaluation indicators that affect shale gas production and productivity; then, each indicator is comprehensively analyzed and evaluated based on the fuzzy transformation principle to determine its weight related to the effects on the evaluation target and its value range; finally, the evaluation and prediction system for the evaluation target is established to provide a theoretical basis and guidance for the quantitative evaluation and prediction of the evaluation target. The main implementation steps of this method are as follows:

(1) Determine $m$ evaluation parameters and divide them into $n$ evaluation units according to shale gas exploration and development demand, thus an evaluation matrix $A$ including $m$ evaluation indicators and $n$ evaluation units is established, as follows.

$$\mathrm{A} = \begin{bmatrix} a_{11} & a_{12} & \cdots & a_{1m} \\ a_{21} & a_{22} & \cdots & a_{2m} \\ \vdots & \vdots & \vdots & \vdots \\ a_{n1} & a_{n2} & \cdots & a_{nm} \end{bmatrix} \tag{6.1}$$

where $a_{i,j}$ is a normalized indicator value; $I = 1, 2, \ldots, m$; $j = 1, 2, \ldots, n$.

(2) Operate matrix A based on the max-min criterion in the optimization algorithm to determine the optimal values of each indicator, that is, the corresponding vectors $G$ and $B$:

$$\mathrm{G} = \begin{pmatrix} \max a_{il}, & \max a_{i2} & ,\ldots, & \max a_{im} \\ 1 \leq i \leq m & 1 \leq i \leq m & & 1 \leq i \leq m \end{pmatrix} = (g_1, g_2, \ldots, g_m) \tag{6.2}$$

$$\mathrm{B} = \begin{pmatrix} \min a_{i}, & \min a_{i2} & ,\ldots, & \min a_{im} \\ 1 \leq i \leq m & 1 \leq i \leq m & & 1 \leq i \leq m \end{pmatrix} = (b, b_2, \ldots, b_m) \tag{6.3}$$

where $g_i$ is the maximum value of the $i$th evaluation indicator, and $b_{i,}$ is the minimum value of the $i$th evaluation indicator.

(3) Determine the weight of each evaluation indicator using the analytic hierarchy process under the constraint of experts' review opinions by combining the shale gas exploration and development experience in the study area.

$$W = (W_1, W_2, \ldots, W_{m,}) \tag{6.4}$$

where $w_i$ is the weight of the $i$th evaluation indicator. A higher $w_i$ is associated with greater effects of $i$th evaluation indicator.

(4) Calculate the comprehensive prediction and evaluation value $V_i$, and prioritize the indicator according to their $V_i$ values.

$$V_i = \frac{1}{\left[\sum_{j=1}^{m} W_j(g_i - a_{ij}) / \sum_{j=1}^{m} W_j(a_{ij} - b_j)\right]^2} \tag{6.5}$$

Equation (6.5) shows that the final evaluation value $V_i$ falls in the range of 0–1, and a larger $V_i$ means that the corresponding area is more favorable to shale gas exploration and exploitation, and vice versa. The final evaluation results can be linked to and calibrated using the production or productivity parameters free from the interference of engineering factors, thus allowing more direct comprehensive prediction and evaluation of the distribution of sweet spots in the study area.

### 6.2.2. *Quantitative prediction system sweet spots*

Seismic technique-based reservoir characterization is the core of shale gas reservoir prediction and evaluation. The specific seismic work includes predicting the burial depth, effective thickness, distribution range, and occurrence of shale reservoirs and determining sweet spots with high organic matter abundance, developed fractures, good physical properties, and high brittle mineral content within favorable areas (Table 6.3).

After determining the data volumes and two-dimensional (2D) distribution of the TOC content, porosity, effective thickness of reservoirs, formation pressure coefficient, and formation brittleness index in the study area, this study selected the data on the test production of two production wells with the same or similar construction parameters and process as the output according to the production conditions on site and used key evaluation parameters obtained through log interpretation (e.g., the TOC content) as the input. Through the automatic optimization analysis using the fuzzy optimization algorithm, the value range and weight of each key prediction and evaluation parameter were finally determined. In this way, this study established a model or system for the quantitative evaluation and prediction of sweet spots in the study area.

Table 6.3 shows that the formation pressure coefficient has the largest weight, followed by the thickness of high-quality reservoirs,

**Table 6.3.** Quantitative seismic prediction system for shale gas sweet spots in the Changning area.

| | | | Evaluation criteria | | |
|---|---|---|---|---|---|
| **Indicator** | **Evaluation parameter** | **Weight (%)** | **Class I sweet spot** | **Class II sweet spot** | **Class III area** |
| 1 | TOC content (%) | 20 | ≥3 | 2–3 | <2 |
| 2 | Thickness of high-quality reservoirs | 22 | ≥15 | 10–15 | <10 |
| 3 | Porosity (%) | 15.5 | ≥5 | 3–5 | <5 |
| 4 | Formation pressure coefficient | 30.5 | ≥1.6 | 1.6–1.2 | <1.2 |
| 5 | Brittleness index (%) | 12 | ≥55 | 40–55 | <40 |

TOC content, porosity, and brittleness index in decreasing order. The former three factors have a significant controlling effect on the distribution of shale gas sweet spots, while the latter two factors have a lower weight because the porosity and the brittleness index of the study area are generally high and continuously distributed.

### 6.2.3. *Quantitative seismic prediction of sweet spots*

Based on the seismic petrophysical analysis, this study obtained the data volumes using the full gather-based simultaneous pre-stack seismic inversion technique and then determined the 2D distribution of each key evaluation parameter. Among them, the TOC content was derived from the inversion of the density data volume and the interpretation using the quantitative interpretation template of the TOC content; the porosity was converted from the P-wave impedance data volume; the thickness of high-quality reservoirs was calculated according to the statistics of the TOC content data volume and the lower limit of the TOC content of the reservoirs in this area; the formation pressure coefficient was calculated using the new formation pressure formula proposed in this chapter with the data volumes of P-wave velocity, S-wave velocity, and density obtained from the pre-stack inversion as input data, and the brittleness index was calculated using the brittleness index formula based on the data volumes of Young's modulus and Poisson's ratio. Figure 6.2 shows the distribution maps of the key parameters of the Long-$I_1$ Submember in the Changning block. Figure 6.2(a) is the 2D distribution map of the average TOC content. According to this figure, the block has TOC content of 2–6% mostly (average: approximately 4%), and the areas with high TOC content are mainly distributed in the central and central-eastern portions of the block. Figure 6.2(b) is the 2D distribution map of the average porosity. This figure shows that the Changning block has average porosity of 2–9% (average: approximately 6%), which is distributed rather consistently. Moreover, the areas with high porosity are mainly distributed in the central and central-eastern portions of the work area. Figure 6.2(c) is the 2D distribution map of the thickness of high-quality reservoirs, which ranges from 5 m to 25 m and shows a preferential distribution in the eastern portion of the study area. Figure 6.2(d) is the 2D distribution map of the average formation pressure coefficient, showing a range of 1.4–1.7 and values above 1.5 in most areas, as well as widely distributed overpressured strata.

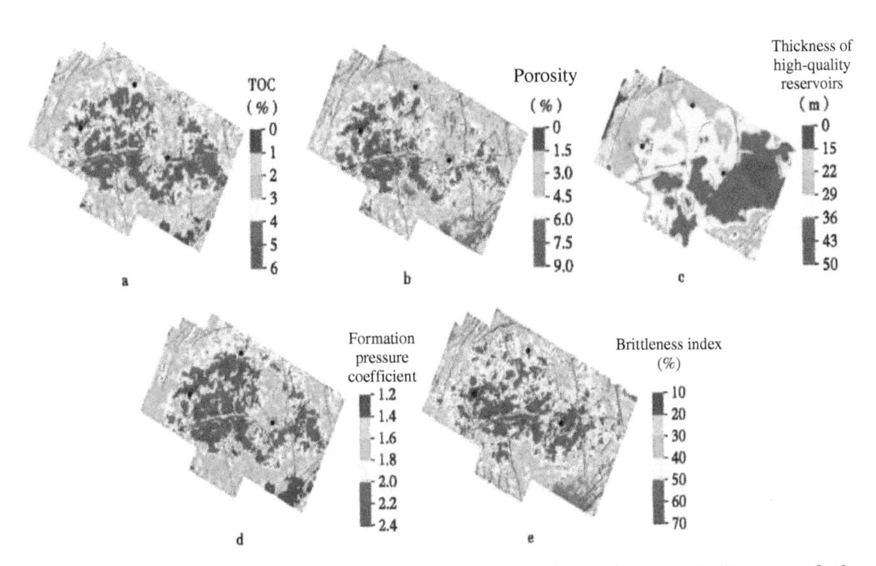


**Figure 6.2.** 2D distribution maps of each single evaluation indicator of the Changning area.

A quantitative comprehensive prediction and evaluation of shale gas sweet spots in the Changning block can be achieved by normalizing the key evaluation parameters using the quantitative seismic prediction system for shale gas sweet spots in the block (Table 6.4) and then adding up their weighted values. The final prediction results were obtained by calibrating the prediction results against the test production of drilled wells, as shown in Figure 6.3. Thanks to this calibration, this method can provide a quantitative prediction of the production of the block on the premise that the construction parameters of the wells selected for calibration (e.g., the length of their horizontal sections and the number of fractured segments) are roughly the same to eliminate the interference of engineering factors as far as possible.

The colors in Figure 6.3 represent the various potential for sweet spots, where areas in red and purple are Class I sweet spots, areas in yellow are Class II sweet spots, and other colors denote non-sweet spots. The Class I sweet spots, with a predicted daily production of more than $20 \times 10^4$ m$^3$ of natural gas, are the most favorable for shale gas development and are recommended to be the priority choice for development. The Class II sweet spots, with a predicted daily production of $(15–20) \times 10^4$ m$^3$ of natural gas, are favorable for shale gas development and are recommended to be the second choice for development. The Class III areas, with a predicted daily

**Table 6.4.** Comparison of the test production and predicted results of verification wells in the Changning block.

| Platform | Well | Predicted results | Test production ($10^4$ m$^3$/d) |
|---|---|---|---|
| H4 | H4-1 | Class I sweet spots | 22.47 |
| | 114-2 | | 23.68 |
| | 114-3 | | 20.56 |
| | 14-4 | | 23.11 |
| | 14-5 | | 17.51 |
| | 14-6 | | 18.2 |
| H2 | 12-2 | Class III area | 3.82 |
| | 12-3 | | 1.00 |
| | 12-5 | | 6.64 |
| | 12-6 | | 4.81 |
| | 12-7 | | 5.66 |

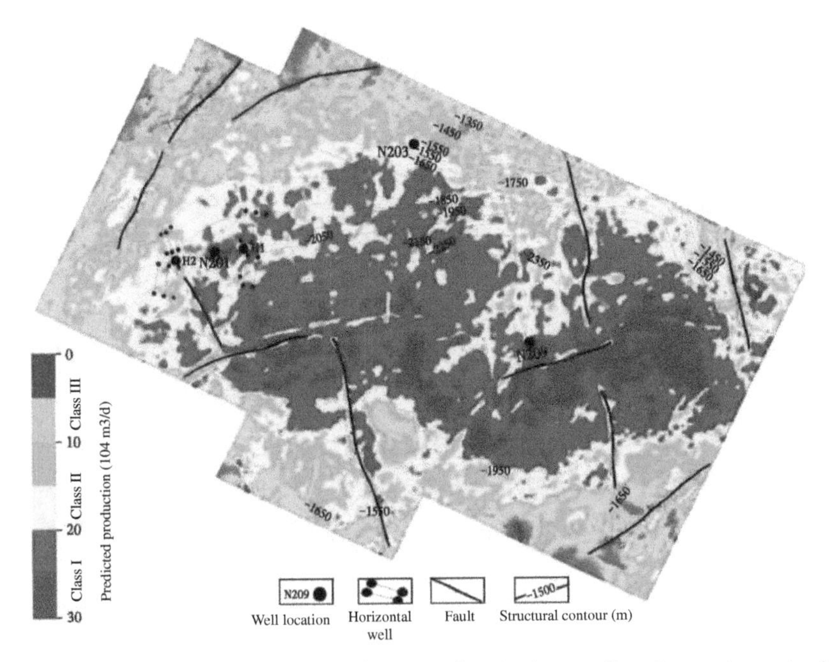


**Figure 6.3.** 2D distribution of comprehensively predicted sweet spots in Changning block.

production of below $15 \times 10^4$ $m^3$ of natural gas and lower values of all key evaluation indicators, are recommended to be developed selectively.

To verify the reliability of the predicted results, the production test results of newly drilled wellpads H4 and H2 in the Changning block were selected to verify and analyze the prediction results (Table 6.4 and Figure 6.3). Since the production results are controlled by both engineering and geological factors, the interference of engineering factors must be eliminated when selecting verification wells. Given that the two wellpads have roughly the same construction parameters and processes, their test production is basically free from the effects of engineering factors and the wells meet the requirements for verification. Moreover, the production results (e.g., the test production) represent the reservoir geological conditions in the areas at the wellpad locations.

The six horizontal wells of wellpad H4 are all drilled in Class I sweet spots as identified through the seismic prediction. The drilling and testing results show that the six horizontal wells are all high-yield industrial gas wells, with daily production during production tests of (17.51–23.68) × $10^4$ m (average: 20.92 × $10^4$ $m^3$).

The five horizontal wells of wellpad H2 are all drilled in Class III areas as identified through the seismic prediction. The drilling and testing results show that the five horizontal wells are all low-yield gas wells, with daily production during the production tests of (1.0–6.64) × $10^4$ $m^3$ (average: 4.39 × $10^4$ $m^3$). Table 6.4 shows the comparison results. The distribution trend of sweet spots identified through the comprehensive quantitative seismic prediction is consistent with that of the production test results. The verification results verify the accuracy of the prediction results and the reliability of the method. Therefore, the prediction results can provide an important basis for the well deployment in future developments.

## 6.3. Summary

(1) This study applied the fuzzy optimization method to the comprehensive quantitative prediction of shale gas sweet spots for the first time. Using this method, this study determined the weights of five key evaluation parameters controlling shale gas production (e.g., TOC content and reservoir pressure) and established a quantitative seismic prediction system for sweet spots in the Changning block for the first time.

(2) Using the established quantitative evaluation system for sweet spots, this study conducted the seismic quantitative prediction of sweet spots in the Changning block based on the results from the petrophysical analysis and pre-stack simultaneous inversion. Then, this study compared the final evaluation results and the test production of newly drilled wells with roughly the same engineering factors. The comparison results show that the quantitative seismic prediction system developed in this study is highly reliable.

(3) Vertically, the high-quality shale reservoirs in the Longmaxi Formation are mainly distributed in the Long-$I_1$ Submember, which is a favorable interval for shale gas development due to its high TOC content, anomalously significant formation overpressure, and high brittleness index. Moreover, the lower intervals of the

Longmaxi Formation display higher key evaluation indicators such as TOC content, formation pressure coefficient, and brittleness index. Laterally, the sweet spots in the formation are mainly distributed in the eastern and central-eastern portions of the study area, which have consistently distributed key indicators, such as thickness and average TOC content.

(4) This study comprehensively evaluated and classified the 2D distribution of shale reservoirs. Class I and Class II sweet spots are mainly distributed in the central and central-eastern portions of the work area and are recommended to be developed preferentially. Class III areas have low values of all key indicators and are recommended to be developed selectively.

## Bibliography

Aliouane L, Ouadfeul SA. Sweet spots discrimination in shale gas reservoirs using seismic and well-logs data: A case study from the Worth basin in the Barnett Shale. *Energy Procedia*, 2014, 59: 22–27.

Bustin AM, Bustin RM. Importance of rock properties on the producibility of gas shales. *International Journal of Coal Geology*, 2012, 103: 132–147.

Chen GH, Bai YH, Chen XZ, *et al.* A new method for comprehensively identifying shale gas sweet spots vertically and its quantitative evaluation. *Acta Petrolei Sinica*, 2016, 37 (11): 1337–1342, 1360.

Chen S, Zhao WZ, Ouyang YL, *et al.* Identification of shale gas sweet spots with the geophysical method — A case study of lower Silurian Longmaxi Formation in Changning area, Sichuan Basin. *Natural Gas Industry*, 2017, 37 (5): 20–30.

Chen S, Zhao WZ, Ouyang YL, *et al.* Prediction of sweet spots in shale reservoir based on geophysical well logging and 3D seismic data: A case study of lower Silurian Longmaxi Formation in W4 block, Sichuan Basin, China. *Energy Exploration & Exploitation*, 2017, 35 (2): 147–171.

Cheng F, Yu G, Zhang YS. Rock physics analysis for shale gas formation "sweet spot" prediction in South China. In *International Geophysical Conference & Exposition*, Beijing, China, 2041, pp. 1145–1148.

Chopra S, Sharma R, Keay J, *et al.* Shale gas reservoir characterization workflows. *Society of Exploration Geophysicists*, 2012, DOI:10.1190/SEGAM 2012-1344.1.

Close D, Perez M, Goodway B, *et al.* Integrated workflows for shale gas and case study results for the Horn River Basin, British Columbia, Canada. *The Leading Edge*, 2012, 31 (5): 556–569.

Deng Y, Chen S, Ouyang YL, *et al.* Comprehensive seismic prediction method of shale gas sweet spots in Weiyuan area of southwest Sichuan and its application. *Petroleum Geology & Oilfield Development in Daqing,* 2019, 38 (2): 112–122.

Ding X, Tan XC, Luo B, *et al.* Multi-parameter carbonate reservoir evaluation based on grey fuzzy theory. *Journal of Southwest Petroleum University,* 2008, 30 (5): 88–92.

Dodds KJ, Dewhurst DN, Siggins AF, *et al.* Experimental and theoretical rock physics research with application to reservoirs, seals and fluid processes. *Journal of Petroleum Science and Engineering,* 2007, 57 (1): 16–36.

Dong DZ, Gao SK, Huang JL, *et al.* On the prospect of shale gas exploration and development in Sichuan Basin. *Natural Gas Industry,* 2014, 34 (12): 1–15.

Hampson DP, Russell BH, Bankhead B. Simultaneous inversion of pre-stack seismic data. In *SEG-2005: 1633-1637, 2005SEG Annual Meeting,* 6–11 November 2005, Houston, Texas, USA.

He XP. Evaluation system of shale gas sweet spots in east Sichuan Basin and influencing factors of enrichment and high yield. *Natural Gas Industry,* 2021, 41 (1): 59–71.

He Y. Quantitative evaluation of reservoirs based on fuzzy comprehensive evaluation and analytic hierarchy process: A case study of Xujiahe Formation in Baojie area. *Petroleum Geology and Recovery Efficiency,* 2011, 18 (1): 23–25, 29.

Hou HX, Ouyang YL, Zeng QC, *et al.* Seismic prediction technology of shale gas sweet spots in Longmaxi Formation in south Sichuan. *Coal Science and Technology,* 2017, 45 (5): 154–163.

Hu BL, Ping WW, Zheng KG, *et al.* Prediction of favorable shale gas areas with GIS-based fuzzy optimization method: A case study of Xiashihezi Formation in Huainan Coalfield. *Fault-Block Oil & Gas Field,* 2015, 22 (2): 189–193.

Huang HD, Ji YZ, Zhang C, *et al.* Application of seismic fluid identification method to prediction of shale gas sweet spots in Sichuan Basin. *Journal of Palaeogeography,* 2013, 15 (5): 672–678.

Jiang TX, Bian XB. New technology of shale gas reservoir evaluation: Sweetness evaluation method. *Petroleum Drilling Techniques,* 2016, 44 (4): 1–6.

Liao DL. Evaluation methods and engineering application of double sweet spots in shale gas reservoirs. *Petroleum Drilling Techniques,* 2020, 48 (4): 94–99.

Peng CZ, Peng J, Chen YH, *et al.* Seismic prediction of shale gas sweet spots in Da'anzhai Section of Yuanba area, Sichuan Basin. *Natural Gas Industry*, 2014, 34 (6): 42–47.

Ross DJK, Bustin RM. Characterizing the shale gas resource potential of Devonian-Mississippian strata in the western Canada sedimentary basin: Application of an integrated formation evaluation. *AAPG Bulletin*, 2008, 92 (1): 87–125.

Sena A, Gastillo C, Chesser K, *et al.* Seismic reservoir characterization in resource shale plays: "Sweet spot" discrimination and, optimization of horizontal well placement. In *2011 Annual International Meeting, SEG.* San Antonio, USA, 9–14, September. Extended Abstract, 2011, pp. 1744–1748.

Shanna RK, Chopra S, Vernengo L, *et al.* Reducing uncertainty in characterization of Vaca Muerta Formation Shale with poststack seismic data. *The Leading Edge*, 2015, 34 (12): 1462–1467.

Sun HQ, Zhou DH, Cai XY, *et al.* Progress and prospect of shale gas development of Sinopec. *China Petroleum Exploration*, 2020, 25 (2): 14–26.

Wang SQ, Wang SY, Man L, *et al.* Appraisal method and key parameters for screening shale gas play. *Journal of Chengdu University of Technology*, 2013, 12 (6): 609–620.

Wang YM, Wang SF, Dong DZ, *et al.* Lithofacies characterization of lower Silurian Longmaxi Formation in south Sichuan. *Earth Science Frontiers*, 2016, 23 (1): 119–133.

Xiong J, Liu XJ, Liang LX, *et al.* Study on the difference of shale reservoirs between the upper and lower members of Longmaxi Formation in Changning area, Sichuan Basin. *Journal of Northwest University*, 2015, 45 (4): 623–630.

Yang RZ, Zhao ZG, Pang HL. Shale gas sweet spots: Geological controlling factors and seismic prediction methods. *Frontiers in Earth Science*, 2012, 19 (5): 339–347.

Zhang JC, Jiang SL, Tang X, *et al.* Accumulation types and resources characteristics of shale gas in China. *Natural Gas Industry*, 2009, 29 (12): 109–114.

Zhang ZW, Cheng YH, Zhao FT. Comprehensive evaluation of shale gas geological characteristics and sweet spots in southwest Sichuan. *China Resources Comprehensive Utilization*, 2019, 37 (8): 43–47, 63.

Zhao WZ, Jia AL, Wei YS, *et al.* Progress and prospect of shale gas exploration and development in China. *China Petroleum Exploration*, 2020, 25 (1): 31–44.

Zhou DH, Jiao FZ. Evaluation and prediction of shale gas sweet spots: A case study of Jurassic system in Jiannan area of Sichuan Basin. *Petroleum Geology & Experiment*, 2012, 34 (2): 109–114.

Zhou YM, Liu XJ, Zhang CH, *et al.* Time-frequency electromagnetic exploration technology and application for quickly identifying shale gas "sweet spot" targets. *Geophysical and Geochemical Exploration*, 2015, 39 (1): 60–63, 83.

https://doi.org/10.1142/9789811283185_0007

# Chapter 7

# Seismic Prediction of Shale Gas Sweet Spots and a Case Study of Seismic–Geological–Engineering Integration

The integrated exploration and development of shale gas is a multidisciplinary systematic engineering. The seismic prediction of shale gas sweet spots is closely linked to other fields such as horizontal well trajectory design, horizontal well drilling, fracturing construction scheme design, and fracturing performance evaluation. Therefore, there is an urgent need to deal with various scientific and technical challenges in shale gas exploration and development by utilizing the multidisciplinary comprehensive advantages of the seismic–geological–engineering integration. This chapter first discusses the significance and study philosophy of the seismic–geological–engineering integration in shale gas exploration and development. Based on the investigation of the Changning block 201, this study introduces the challenges, research approaches, and technical routes of the seismic–geological–engineering integration, the integrated seismic–geological modeling, the horizontal well geosteering, and the design and optimization of the hydraulic fracturing construction scheme, providing a successful case for the seismic–geological–engineering integration.

## 7.1. The seismic–geological–engineering integration in the shale gas exploration and development

### 7.1.1. *Significance of the seismic–geological–engineering integration*

Vastly different from conventional hydrocarbon production, shale gas exploration and development are characterized by a distinctive integrated exploration and development. Therefore, it is necessary to build a multidisciplinary collaborative work team and platform, establish the workflows and processes of the geology–engineering integration, and provide integrated geology–engineering solutions to improve and optimize shale gas development with scale efficiency. The application of seismic techniques and data, which are used throughout the shale gas exploration and development, play an important role in many aspects, such as the selection and evaluation of shale gas exploration areas, the prediction of reservoir sweet spots, well deployment, the design and adjustment of well trajectories, the design of fracturing construction schemes, fracturing performance evaluation, and the adjustment of fracturing construction schemes. Therefore, the geology–engineering integration is essentially the seismic–geological–engineering integration.

### 7.1.2. *Research philosophy of the seismic–geological–engineering integration*

This study investigates the seismic–geological–engineering integration in the shale gas exploration and development on the basic principle of hierarchy from blocks to wellpad areas, as follows: (1) it conducts the well logging and the geophysical and geological study of the whole block to clarify its general structure, its reservoir properties, and the distribution patterns of its natural fractures; (2) it selects a typical wellpad by keeping pace with the drilling program and reservoir stimulation based on the block model, and then establishes fine-scale geological and mechanical models by integrating multidisciplinary and multi-source data with seismic data as the carrier; (3) it applies the study results to engineering based on the integrated geological model, thus providing support for well drilling and completion; (4) it selects a wellpad with complete microseismic monitoring and log data to analyze the fracturing performance and simulate hydraulic fractures, and further analyze the main factors affecting hydraulic fractures and the influence of fracture morphology on production capacity; (5) it tracks the shale gas testing and flowback of key wells, formulates a short-term pilot production plan, evaluates the single-well production capacity, and develops reasonable production regimes. The specific steps include:

(1) Building the geophysical and geological models of the shale gas reservoirs in the study area, including:
   - the petrophysical analysis and log evaluation of reservoirs;
   - the interpretation of seismic data and the seismic prediction of the key evaluation parameters and geomechanical parameters of reservoirs; and
   - geological modeling of the reservoirs.

(2) Establishing a fine-scale model of a typical wellpad area, including building:
   - a precise geological model of the wellpad area;
   - a one-dimensional (1D) precise geomechanical mod;
   - a three-dimensional (3D) pore pressure model of the wellpad area; and
   - a 3D geomechanical model of the wellpad area.

(3) Providing support for seismic geosteering and fracturing engineering, including:
   - the seismic–geological integrated design of horizontal well trajectories and horizontal well geosteering;
   - the optimization and adjustment of horizontal well trajectories; and
   - the design and real-time adjustment of fracturing construction schemes.

(4) Evaluating fracturing performance, including:
   - parameter fitting of the fracturing pumping process;
   - the simulation of the fracture network morphology based on microseismic monitoring;
   - seismic–geological–engineering integrated estimation of effective SRV and recoverable reserves;

- Production dynamics prediction;
- The optimization of reservoir stimulation parameters based on the prediction of seismic–geomechanical parameters;
- The optimization of production horizons and well spacing.

(5) Engineering analysis of gas reservoirs at key well locations.

## 7.2. Performance of the seismic–geological–engineering integration in the Ning-201 block

### 7.2.1. *Geological setting*

The area of well Ning-201 falls within the Changning–Weiyuan National Shale Gas Demonstration Area and is the area where the great performance of seismic–geological–engineering integration has been achieved using relatively mature techniques. This area is located in the southwestern Sichuan Basin, spans across the Changning, Gong, Xingwen, and Junlian counties in Yibin City, Sichuan Province, and is a part of the Shuifu–Xuyong mining right area. It has a typical mountainous terrain dominated by medium and low mountains and hills, with an elevation of 400–1,300 m.

In terms of structural position, the area of well Ning-201 lies in the the low and steep structural part of the uplift in the central part of the paleo-depression in southern Sichuan and the syncline zone of the south limb of the Changning anticline of the Loushan fold belt. The target layers for shale gas exploration and development mainly include the organic-rich shale intervals in the Upper Ordovician Wufeng Formation and the lower part of Lower Silurian Longmaxi Formation, with a total thickness of 30–50 m and depth of 2,300–3,200 m.

The Wufeng–Longmaxi Formations are shelf-facies sediments, forming organic-rich mudstones characterized by simple lithology, fine-grained particles, thick beds, wide lateral continuity, and abundant biological fossils. The shelf-facies sedimentary strata can be subdivided into two subfacies, that is inner shelf and outer shelf, and seven microfacies such as organic-rich siliceous mud shelf microfacies and organic-rich silty mud shelf microfacies. The Wufeng Formation and the Long-$I_1$ Submember have the aforementioned two sedimentary microfacies, which are the most favorable for the accumulation and high production of shale gas.

The shale gas reservoirs of the Wufeng–Longmaxi Formations mainly consist of black carbonaceous shales, black shales, siliceous shales, black mudstones, and black and grayish-black silty mudstones. These reservoirs have high brittle mineral content, with an average of over 70%. The brittle minerals are dominated by siliceous minerals and thus have high compressibility. The clay minerals mainly consist of illite (52.2%), chlorite (25%), and illite/montmorillonite (I/M) interstratified minerals, with a low amount of expansive minerals. The organic matter mainly consists of sapropel and has kerogen of type I and $R_0$ generally above 2.5%, thus reaching the over-mature stage and mainly producing dry gas. The Wufeng Formation–Long-$I_1$ Submember interval has high organic matter content overall, with measured total organic carbon (TOC) content of 3.0–4.2% and a log-derived TOC content of 2.7–4.5%.

The maximum horizontal principal stress of the shale gas reservoir interval is mainly in a SE-EW direction and in an NE direction locally. The Wufeng Formation–Long-$I_1$ Submember interval has triaxial compressive strength measured on core materials of 181.73–321.74 Mpa (average: 254.04 Mpa), Young's modulus of $(1.548–5.599)\times10^4$ Mpa (average: $3.52 \times 10^4$ Mpa), and Poisson's ratio of 0.158–0.331 (average: 0.225).

There are various types of reservoir spaces in the Wufeng Formation–Long-$I_1$ Submember interval in the Changning block, mainly including pores and micro-fractures. The pores consist of organic matter-hosted pores, intercrystalline pores, intracrystalline dissolution pores, and intergranular pores. The fractures include tectonic fractures, diagenetic fractures, dissolution fractures, and hydrocarbon-generating fractures. The interval has high porosity, with measured values of 2.0–6.8% and log-derived values of 3.6–7.3%. It has low matrix permeability of $(0.714–1.48)\times10^4$ mD, with an average of $1.02 \times 10^4$ mD. Moreover, there is a weak correlation between the porosity and permeability of the interval (nitrogen determination method), and the bedding fractures and micro-fractures of shales have significant effects on shale permeability.

The shale gas has a methane content of over 97%, small amounts of hydrocarbons above $C_3$ (no hydrogen sulfide), and $CO_2$ content of 0.22–0.54%. This highly mature natural gas has a drying coefficient

($C_1/C_{2+}$+) of 134.65–282.98, high gas saturation of 50–70%, and high total gas content, with measured values of 2.0–3.5 $m^3/t$ and log-calculated values of 2.9–7.4 $m^3/t$ (average: 4.8 $m^3/t$).

The area of well Ning-201 is located in a relatively stable tectonic region, with the upper and lower Longmaxi strata serving as well-preserved tight sealing layers. Higher TOC content corresponds to favorable gas-bearing properties. It is currently considered that the organic matter abundance of the area of well Ning-201 is a key factor controlling the gas-bearing property of the shales in the Wufeng–Longmaxi Formations.

Regarding drilling engineering design, the area of well Ning-201, after preliminary development tests, saw the adoption of a single wellpad with six wells in a double-row layout, whose technologies are roughly mature. The casing program is based on a "three-section and three-completion" design. The well trajectories adopt the design of "vertical section — building up section — holding interval — building up section (direction correction) — horizontal section".

A completion method with bridge plugs and wireline cluster perforations is adopted in the area of well Ning-201. At present, the cluster spacing is generally determined on the principle of forming multiple fractures and increasing the complexity of fractures by utilizing the stress interference between clusters. By combining the actual well construction, the cluster spacing in the area is determined to be 15–20 m. The level spacing is determined on the principle of effective stimulation of reservoirs within sections, smooth pumping of pipe strings, and an optimal number of perforation clusters. By comprehensively considering the pumping of the pipe strings, the level spacing of stimulated wells, and microseismic monitoring, three perforation clusters per section are mainly adopted, and the level spacing is determined to be 60–80 m corresponding to the cluster spacing of 15–20 m. The pumping displacement is determined on the principles of ensuring successful construction, increasing the net pressure, and maintaining the integrity of the wellbore. Increasing the pumping rate is conducive to the increase in the net pressure in the fractures. Given that the regional horizontal stress difference is approximately 13 Mpa, the general pumping displacement is determined to be 12–14 $m^3$/min based on the pumping displacement that can be achieved under the control pressure as well as the simulation results and the analysis of the net pressure in the fractures of the stimulated wells. According to the pressure response of adjacent wells and the well testing results of interference on site, approximately 1,800 $m^3$ of fluids per section were adopted under the tunnel spacing of horizontal wells of 400–500 m. The slug injection-type sand addition method is adopted. The sand amount for a single section is generally 80–120 tons according to the limitation on liquid volume and the maximum sand concentration as well as the actual sand addition parameters in the early stage.

In terms of the drainage and production strategies, the early post shut-ins show that there is a certain correlation between the pressure drop during the shut-in period and the test production and that the post shut-in is conducive to further expansion of fractures. To further explore the effects of post shut-in on the stimulation performance, the drainage and production regime of post shut-in for 3–5 days followed by opening for drainage with a 3–4 mm nozzle is still adopted at present. After the bottomhole pressure is reduced to the closure pressure of the reservoirs, the size of the nozzle is adjusted to ensure stable and continuous drainage. The reservoir energy is sufficient for liquid production in the early stage of gas well production.

### 7.2.2. *Major challenges of the seismic–geological–engineering integration*

With the progress in the shale gas exploration and development in southern Sichuan, there is an urgent need to deal with the following challenges:

(1) **The prediction of engineering factors concerned in the shale gas development:** Engineering factors (e.g., rock elastic parameters, *in situ* stress, and pore pressure) are especially a concern in shale gas development. The prediction of the 3D distribution of engineering factors based on seismic and log data is of great significance for the study of wellbore integrity and the optimization of fracturing design.

(2) **Assuring wellbore integrity and the probability of drilling high-quality reservoirs in the horizontal well interval:** Shale gas development faces the challenge of wellbore

integrity due to the complex geological conditions of reservoirs and the strong heterogeneity of *in situ* stress. Meanwhile, it is necessary to ensure smooth well trajectories and the probability of penetration of high-quality reservoirs to ensure construction efficiency and single-well production. However, there are uncertainties in geosteering using a non-azimuthal gamma ray (GR) log due to the cyclical characteristics of shales. Thus, a reliable workflow for drilling and geosteering is required.

(3) **The evaluation of volume fracturing performance and the optimization of fracturing parameters:** The presence of natural fracture systems causes complex microseismic monitoring results, posing a challenge to the evaluation of the effective stimulated reservoir volume. However, the formation mechanisms and control factors of complex fracture networks have seldom been reported, and it is necessary to further study the morphology of hydraulic fractures and the relationships between hydraulic fractures and natural fractures. Therefore, it is necessary to evaluate the stimulation performance, maximize the value of microseismic data, and optimize the fracturing parameters of subsequent wells based on the achievements.

(4) **The optimization of both the drainage and production schemes and the production regimes:** The development of several shale gas fields in North America indicates that shales are highly sensitive to stress, and proper pressure drop management can help maintain medium-long-term production capacity. Therefore, the optimization of the production regime by combining the actual shale reservoir characteristics in the area of well Ning-201 is of great significance for increasing and stabilizing the production of shale gas.

(5) **The understanding of the shale gas development characteristics and the optimization of development-related technical policies:** First, more precise reservoir characterization requires integrated modeling based on geological and geomechanical characteristics. Second, it is necessary to quantify the flow characteristics of shale gas reservoirs. Third, the optimization of the macro development-related technical policies (such as the azimuth, length, and spacing of horizontal wells) is also a concern in the development of shale gas fields.

### 7.2.3. *Research philosophy and technical routes of the seismic–geological–engineering integration*

This study aims to increase production and efficiency. It closely combined multiple disciplines such as geology, logging, seismic, geomechanics, and reservoir engineering and fully utilized the respective advantages of Schlumberger and oil fields. Moreover, by closely combining the comprehensive geological studies with engineering practices, this study achieved real-time decision-making, rapid response, and smoothly rolling operations. It took the initiative to adopt new techniques and methods to strive for new understandings and breakthroughs. The workflows of this study are shown in Figure 7.1.

The seismic–geological–engineering integration for shale gas development requires additional geophysical, geological, and engineering techniques. Many new and advanced techniques have been applied to the integration, involving many aspects, such as the processing and interpretation of seismic data, log interpretation and evaluation, geological modeling, horizontal well geosteering, seismic geomechanics, fracturing performance evaluation, and reservoir stimulation and optimization.

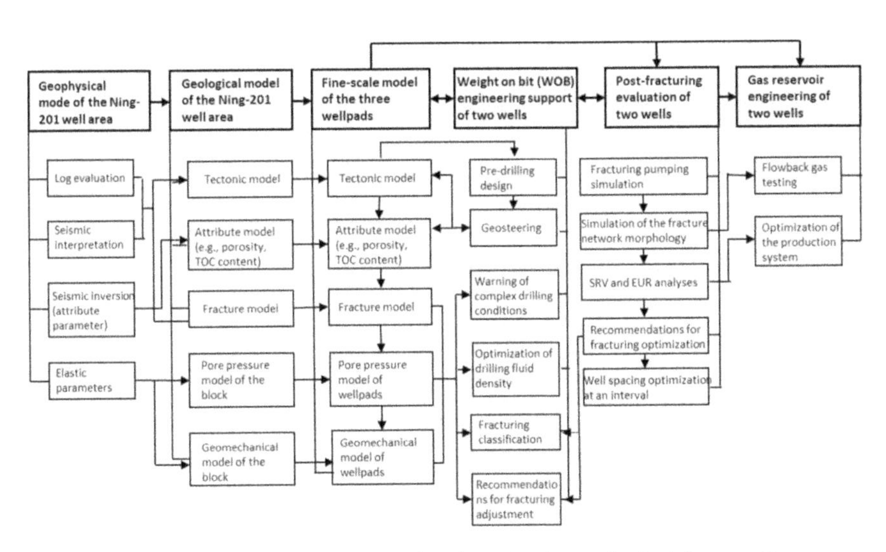


**Figure 7.1.** Technical routes for the seismic–geological integration.

#### 7.2.3.1. *Interpretation of 3D seismic data*

The interpretation of seismic data involves many aspects, including the interpretation of structures and lithology, the prediction of shale reservoir properties (e.g., physical properties and TOC content), the prediction of reservoir engineering factors (e.g., brittleness, *in situ* stress, and fractures), and fluid detection. As for this study, the seismic interpretation mainly includes (1) interpreting the structural horizons of the high-quality shales in the Longmaxi Formation by combining automatic tracking with manual interpretation; (2) interpreting faults and micro-faults using seismic attributes such as edge detection attributes, variance volumes, and ant-tracking attributes. The key techniques include:

(i) Log-seismic calibration of the synthetic seismogram, which is used to determine the reflection characteristics of the Longmaxi Formation in the main target layers;
(ii) Extraction of edge detection attributes, variance volumes, and ant-tracking attributes, which is used to understand the distribution and characteristics of faults in the study area and assist the manual interpretation of faults and micro-faults;
(iii) Automatic and manual horizon interpretation, which is used to ensure the interpretation accuracy of micro-amplitude structures;
(iv) Establishment of a space-variant velocity field based on the horizon interpretation and the log-seismic calibration, which is used to perform the time-to-depth conversion.

#### 7.2.3.2. *Pre-stack simultaneous amplitude versus offset inversion*

Inversion is the main method for the seismic prediction of key geological evaluation parameters and engineering evaluation parameters of shale gas reservoirs. The pre-stack synchronous amplitude versus offset (AVO) inversion is the most critical technique and involves many key technical links.

(1) **Preprocessing of seismic data:** Refers to the necessary preprocessing of seismic data, such as residual moveout correction, processing for improving resolution, multiple attenuations, and

noise suppression. They are conducted to make seismic data applicable to the pre-stack seismic inversion.

(2) **Seismic petrophysical analysis and modeling:** First, edit and conduct control log curves to remove anomalies and ensure that the physical properties meet the theoretical limits through quality control; then, investigate the effects of the compaction, lithology, porosity, and fluids in strata on acoustic attributes; third, define the lithological categories and perform error analysis and control on log and seismic inversion scales; finally, establish the quantitative relationships between the evaluation parameters of shale gas sweet spots and the elastic parameters such as P-wave impedance, S-wave impedance, and density through petrophysical modeling, aiming to provide a basis for the seismic prediction of these key evaluation parameters.

(3) **Pre-stack synchronous AVO inversion:** Stack the pre-stack gathers at different angles based on the AVO forward modeling results, and then directly invert the high-resolution elastic parameters (e.g., P-wave impedance, the $v_p/v_s$ ratio, and density) using the angle stacks. These elastic parameters can be converted into reservoir rock attributes such as lithology, physical properties, and saturations using a petrophysical template established at the petrophysical modeling stage.

(4) **Lithological classification based on LithoCube:** Obtain the probability density function by analyzing log data-derived elastic parameters (e.g., P-wave impedance, $v_p/v_s$) and physical property parameters (e.g., shale content, porosity, and water saturation); determine the lithological classification and the probability of each lithological class by applying the probability density function to pre-stack seismic inversion results; analyze the physical property parameters and the rock elastic parameters, and then determine the relationships between rock elastic parameters and pores and fluids based on the log data of the physical properties of the studied wells. These relationships will be applied to the pre-stack seismic inversion results to generate equiprobable attributes such as porosity and TOC content.

#### 7.2.3.3. *Comprehensive evaluation of shale gas reservoirs*

Establish an accurate log interpretation model for reservoirs based on the advanced techniques for the acquisition, processing, and interpretation of log data; quantitatively calculate the parameters denoting the shale reservoir quality, including shale content, TOC content, porosity, and gas content; comprehensively evaluate the shale gas reservoirs and determine the vertical distribution of shale gas sweet spots. The comprehensive evaluation of shale gas reservoirs will also provide a solid and reliable basis for structural, geological, natural-fracture, and geomechanical modeling, the analysis and evaluation of well drilling and completion strategies, the design of fracturing construction scheme, and the follow-up fracturing performance evaluation.

#### 7.2.3.4. *3D seismic–geological modeling*

To research the reservoir quality and well-completion quality of the study area, various attribute models have been established using 3D geological modeling, which includes the following three techniques.

(1) **Fine-scale structural modeling based on log-seismic integration:** Establish a fine-scale 3D structural model by combining singh-well structural information (e.g., the structural dip angle from image logs and the fine-scale sublayer correlation in the true stratigraphic thickness (TST) domain) with the stratigraphic boundaries obtained from seismic interpretation.
(2) **Attribute modeling based on log-seismic integration:** Establish attribute models reflecting the reservoir quality (e.g., TOC content, porosity, saturation, and gas content) using geostatistical methods based on core analysis data, special log data, and seismic attributes (e.g., seismic inversion).
(3) **Fracture modeling based on multi-scale information:** Analyze and predict fractures from individual wells to well periphery, and then the entire block and build a 3D fracture model by fully utilizing image logs data, microseismic monitoring data, and seismic attributes.

#### 7.2.3.5. *Real-time seismic geosteering of horizontal wells*

The geosteering of horizontal wells is closely combined with the 3D seismic–geological modeling to ensure the geosteering performance of the horizontal well sections and maximize the probability of penetration of shale gas sweet spots. According to the drilling stages of horizontal wells, technical support is provided for the geosteering of horizontal wells in the following three stages:

(1) **Pre-drilling stage:** Establish a pre-drilling geological model to serve the pre-drilling design of horizontal wells and to provide the targeted geosteering recommendations and warnings about complex conditions according to the structural and seismic fracture prediction results of the horizontal well sections.
(2) **Drilling stage**: Assist geosteering engineers in building a real-time geosteering model based on the real-time drilling performance and information obtained through logging while drilling; provide targeted recommendations for real-time geosteering through multi-dimensional interaction (e.g., sections, 3D, and TST domain) and risk control strategies based on the high-accuracy seismic data.
(3) **Drilling completion stage**: Update the models such as the 3D structural model and attribute models according to data on well drilling completion and logs, thus improving the precision of the 3D geological model.

#### 7.2.3.6. *Geomechanical modeling*

Reservoir stimulation is one of the most important techniques for shale gas development. Therefore, an accurate geomechanical model is of great significance to the design of a fracturing construction scheme. Geomechanical modeling includes 1D and 3D modeling.

(1) **Precise 1D geomechanical modeling:** This technique was first proposed by Schlumberger, who first proposed the concept of the mechanical earth model in 2000 as well as the practical methods and processes to improve the precision and reliability of the model. This method and process have been widely applied in many countries and have been accepted and recognized in the industry.
(2) **3D geomechanical modeling:** Based on the geomechanical parameters obtained from seismic inversion, this technique can be applied to precisely characterize the distribution of the stress field in complex geological structures through a 3D finite element simulation.

Through 3D geomechanical modeling, the geomechanical simulation system for oil and gas reservoirs can be used to carry out in-depth analysis, evaluation, and study of the geomechanics-related engineering concerns (e.g., drilling, completion, and fracturing) on different scales from a single well, to a wellpad, then a block, and finally gas fields.

#### 7.2.3.7. *Drilling process evaluation and optimization*

The 1D and 3D geomechanical models are combined with other geoscientific models (e.g., structural models, geological models, attribute models, and natural fracture models) to fully evaluate the effects of *in situ* stress, stress heterogeneity, and stress anisotropy along wellbores, near wellbores, and in the far field. The purpose is to provide more reliable evaluations and prediction methods for well trajectory design, the optimization of drilling fluid density, safe drilling, geosteering for improving production capacity, and wellbore integrity.

#### 7.2.3.8. *Simulation of hydraulic fractures*

The simulation of hydraulic fractures adopts the relevant modules of the fracturing stimulation design software to seamlessly connect with the geophysical, log, geological, and drilling and completion data. It makes full use of both the 3D natural-fracture geomechanical models to precisely simulate the complex fracture network morphology formed by hydraulic fracturing. This technique can simulate fracture propagation and proppant transport considering reservoir heterogeneity, stress anisotropy, and stress shadow effects. Moreover, it can simulate the interactions between hydraulic fractures and natural fractures considering the actual occurrence of natural fractures.

During the simulation of a complex fracture network formed by hydraulic fracturing, the morphology of the generated hydraulic fracture network is fitted with the actual downhole microseismic event points by combining microseismic monitoring data in order to make the simulation results of the hydraulic fracture network consistent with the distribution of microseismic event points, thus improving the precision for the simulation of the hydraulic fracture network.

#### 7.2.3.9. *Evaluation and optimization of fracturing process and scheme*

The numerically simulated complex fracture network system of hydraulic and natural fractures is used to build an unstructured production grid model. Very fine grid cells are used to describe the morphology of the hydraulic fracture network, and the grid permeability is automatically calculated according to the distribution and conductivity of proppant in the hydraulic fracture network. The purpose is to provide a model basis for the numerical simulation of fractured oil reservoirs, realize the multiphase flow simulation of the complex fracture network formed by hydraulic fracturing, form seamless data connection from fracturing to production, and finally establish the optimization workflow from completion fracturing design to production simulation.

#### 7.2.3.10. *Microseismic monitoring and interpretation*

Microseismic monitoring plays an important role in the development of unconventional hydrocarbons such as shale gas. This technique can be used to provide the azimuth and geometry (e.g., length, width, and height) of hydraulic fractures and analyze the deformation amplitude of strata based on the cumulative seismic moment. Regardless of the locations of monitoring wells, this technique can directly reflect the deformation of strata and the spatial-temporal changes in the microseismic response and eliminate the effects of fracturing fluid filtration loss and secondary fractures, thus providing a more precise geometry of the main hydraulic fractures. Moreover, this technique can analyze the complexity of hydraulic fractures according to the ratio of P-wave to S-wave amplitude of the microseismic events and determine the activation of natural fractures and faults during the fracturing process and its impact on fracturing according to the seismic magnitude, thus providing corrections and feedback for natural fracture models and geomechanical models and providing support for the simulation of hydraulic fractures to optimize the fracturing design.

#### 7.2.3.11. *Production capacity evaluation and development scheme optimization*

Based on the integrated software platform, this technique is used to conduct production capacity evaluation and development strategy optimization through the study of gas reservoir engineering. This technique mainly involves:

(1) **Dynamic fitting of single-well production history:** Correct the static model and engineering data using numerical simulation based on single-well production history.
(2) **Prediction of single-well production capacity:** Quantitatively characterize matrix and fractures and establish a single-well geological model through dynamic and static combination; predict the production under different production conditions using numerical simulation combined with engineering knowledge.
(3) **Development scheme optimization:** Predict shale gas sweet spots based on geophysical and geological models and provide the optimization recommendations (e.g., wellpad location and horizon selection) by combining numerical simulation methods during the deployment and design of new wellpads.

### 7.2.4. *Integrated seismic–geological modeling of shale gas reservoirs*

The integrated geoscientific model is a 3D shared geological model that integrates various elements such as structures, reservoir attributes, natural fractures, and geomechanical attributes and can be updated in time and dynamically in the process of the geology–engineering integrated operation. It is the basis for the subsequent optimization of fracturing parameters, landing horizons, production regimes, and technical development policies. It can be applied to engineering such as real-time geosteering, warning about drilling risks, and the real-time adjustment of fracturing parameters.

A 3D shared geoscientific model can be built as per the workflow in the following steps, as shown in Figure 7.2:

(1) Conduct fine-scale single-well log interpretation and multi-well stratigraphic correlation to understand the stratigraphic characteristics, lithological associations, the physical property characteristics of reservoirs, hydrocarbon saturations, petrophysical characteristics in the study area as well as their interwell lateral variations.

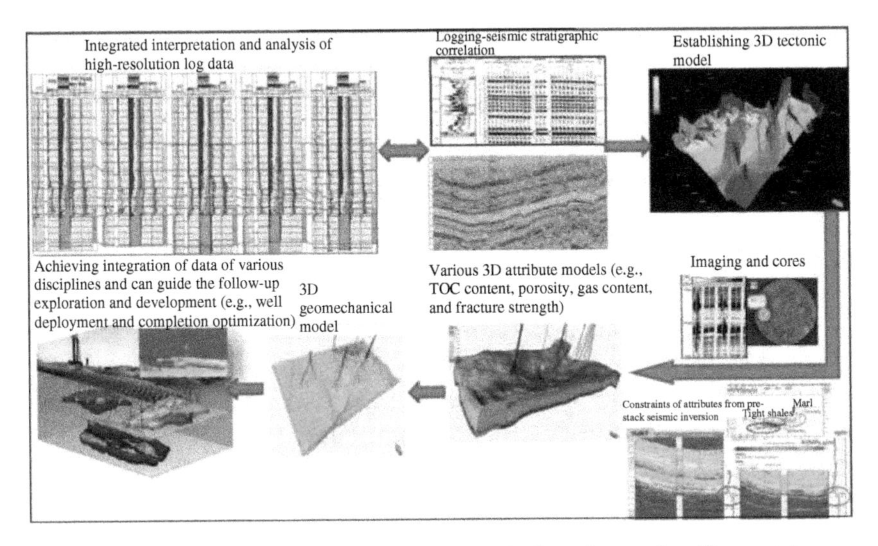


**Figure 7.2.** Workflow for building a 3D shared geoscientific model.

(2) Conduct horizon calibration based on synthetic seismograms to determine the reflection characteristics of key horizons. Multi-well horizon calibrations can also be used to verify the rationality and accuracy of single-well stratigraphic divisions. The relatively accurate stratigraphic division and seismic interpretations can be finally determined through repeated interactive verification using seismic interpretation and stratigraphic correlation.
(3) Conduct pre-stack seismic inversion using the pre-stack seismic gathers and single-well petrophysical analysis results and obtain the 3D distribution of the physical properties and elastic parameters of reservoirs; establish the structural model of the study area using the horizons and faults obtained from seismic interpretation combined with the stratigraphic division results
(4) Establish 3D reservoir attribute models based on the structural model as well as single-well reservoir attributes and the pre-stack seismic inversion results; develop a 3D natural fracture model by combining the single-well natural fracture interpretation and the seismic attributes closely related to natural fractures.
(5) Convert the 1D geomechanical model into a 3D geomechanical model based on the geological attribute models and the natural fracture model; finally, build a 3D shared geoscientific model by

integrating the data of various disciplines to guide the follow-up shale gas exploration and development. All these steps are detailed as follows.

#### 7.2.4.1. *Log interpretation and comprehensive reservoir evaluation*

The log evaluation is researched at three levels:

(1) Verify the consistency between the log interpretation and the core analysis results of pilot holes and establish an accurate log interpretation model after being calibrated using core data, thus ensuring the accuracy of log evaluation.
(2) Correct and analyze the log data of horizontal wells from different service providers to identify the reference horizon for normalization and to conduct deviation correction by making references to the log curves of adjacent pilot holes, i.e., indirect normalization.
(3) Input the log data of horizontal wells after being correlated in a united manner into the log interpretation model of the pilot holes and then carry out acomprehensive log interpretation and evaluation of horizontal wells, thus providing an accurate single-well data basis for subsequent geological and geomechanical modeling. The specific technical routes are shown in Figure 7.3.

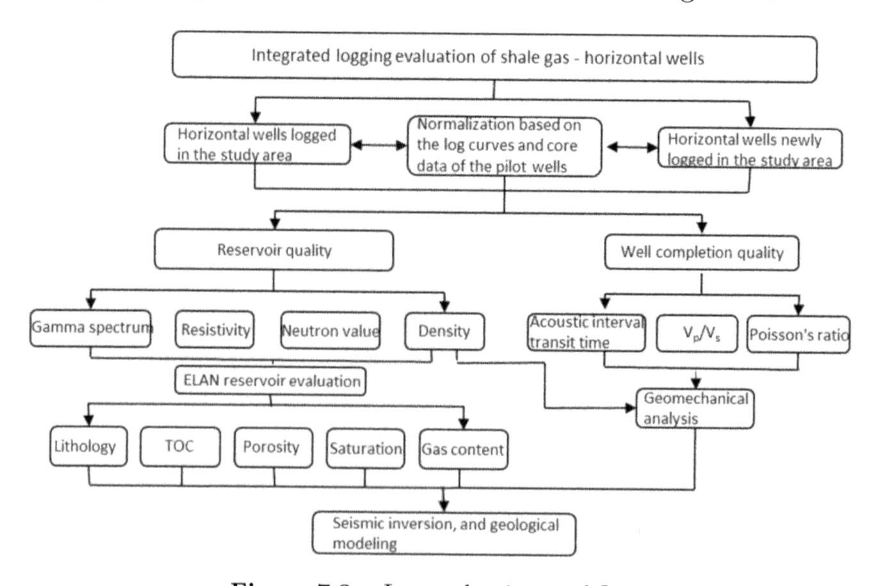


**Figure 7.3.** Log evaluation workflow.

The main steps of log evaluation include:

(1) Conduct core analysis and calibration for the log curves and reservoir evaluation results of the pilot holes in the study area, aiming to reduce log evaluation errors and ensure the accuracy of the log evaluation results; establish an accurate log interpretation model of shale reservoirs and quantitatively calculate the parameters related to reservoir quality, including shale content, TOC content, porosity, and gas content.
(2) Sort out the log data of horizontal wells in the study area and normalize the log curves of the horizontal wells; perform the comprehensive evaluation of reservoirs at the locations of horizontal wells using the log interpretation model built in the first step, thus ensuring the consistency of the survey and interpretation between horizontal and pilot holes.

The comprehensive log evaluation is expected to provide consistent interwell log evaluation and the parameters related to reservoir quality.

Figures 7.4 and 7.5 show the comparison of the log evaluation results before and after being calibrated using core data of wells Ning-201 and Ning-203, respectively. The curves before calibration are the log evaluation results on the well completion, while the curves after calibration are the secondary log evaluation results obtained using the log interpretation model that is calculated using the data from core analysis and testing. After being calibrated using core data, the relative errors of the log evaluation results can be reduced by approximately 5–25%, especially for the calcareous mineral content, porosity, and TOC content.

The log interpretation results of the three pilot holes in the study area were re-corrected using the data from core analysis and testing. Accordingly, the log interpretation model based on core testing was built to ensure reliability.

The log curves of horizontal wells are more susceptible to environmental factors such as well conditions. Moreover, there are different logging service providers, and different logging instruments are used in the study area, leading to certain deviations between log curves. Therefore, quality control and normalization must be carried out for log curves before the log evaluation.

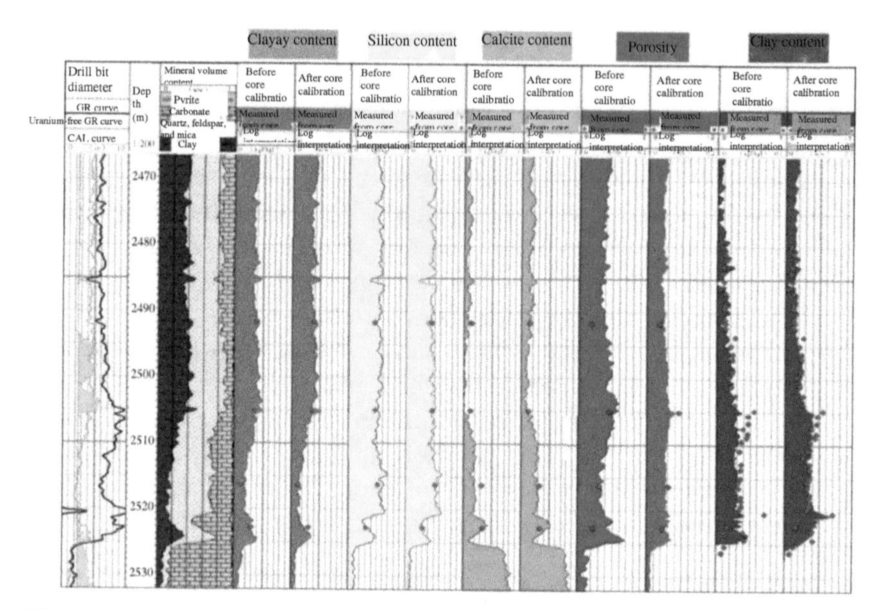


**Figure 7.4.** Comparison diagram of log evaluation results of well Ning-201 before and after being calibrated using core data.

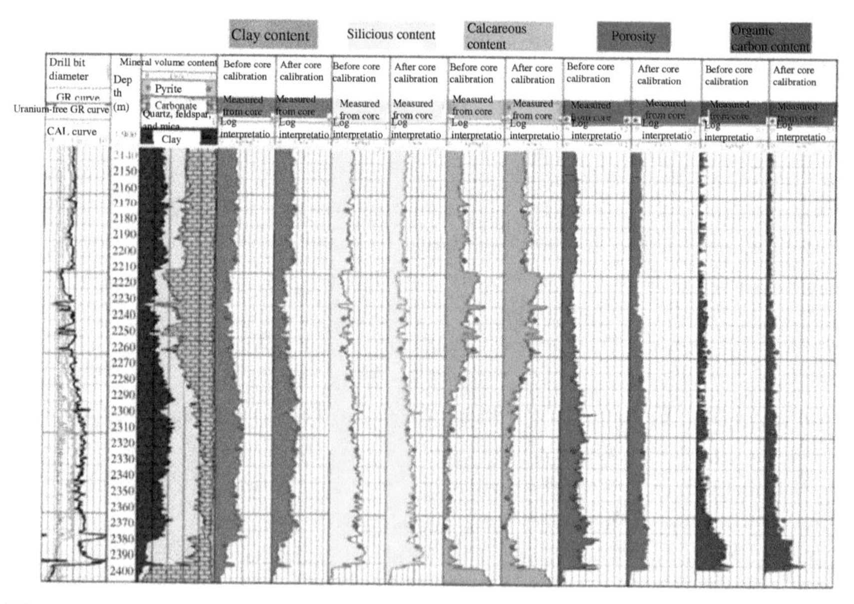


**Figure 7.5.** Comparison diagram of well Ning-203 before and after being calibrated using core data.

Figures 7.6 and 7.7 show the comprehensive log evaluation of wells Ning-201, Ning-203, and Ning-209 and the lateral correlation of parameters related to reservoir quality of these wells, respectively. The high-quality reservoir interval (Wufeng Formation–Long-$I_1^3$ Sublayer) at the three pilot holes in the study area has different thicknesses, which is thinner for the well Ning-209 and similar for wells Ning-201 and Ning-203. The vertical comparison of the shale reservoir parameters (porosity, TOC content, gas content, and brittle mineral content) shows that the Long-$I_1^1$ Sublayer has the most favorable parameters, followed by the Long-$I_1^2$ and Long-$I_1^3$ Sublayers and the Wufeng Formation, and the Long-$I_1^4$ Sublayer has less favorable reservoir parameters. The lateral comparison results show that the three pilot holes have similar parameters related to reservoir quality, with parameters of well Ning-201 being slightly more favorable than those of the other two wells.

#### 7.2.4.2. *Fine-scale seismic interpretation and fracture prediction*

Log data have high longitudinal resolutions, and thus the longitudinal heterogeneity of shale gas reservoirs at the well locations can be obtained through the comprehensive log study. However, it is necessary to use the seismic data to obtain the interwell variation of shale gas reservoirs' properties (i.e., the lateral variation). In other words, it is necessary to predict the lateral variation trends of shale gas reservoirs and the distribution of natural fractures through the structural interpretation of seismic data, the extraction and analysis of seismic attributes, and pre-stack seismic inversion in order to provide a basis for well deployment and guidance for geological modeling.

The research philosophy and technical routes of the seismic interpretation of shale gas reservoirs are basically the same as the conventional interpretation process. The interpretation process of the main horizons is summarized as follows:

(1) Regarding the calibration of key wells, determine the horizons to be interpreted first, and then determine the relationships between the stratigraphic division of the wells and seismic reflection interfaces;

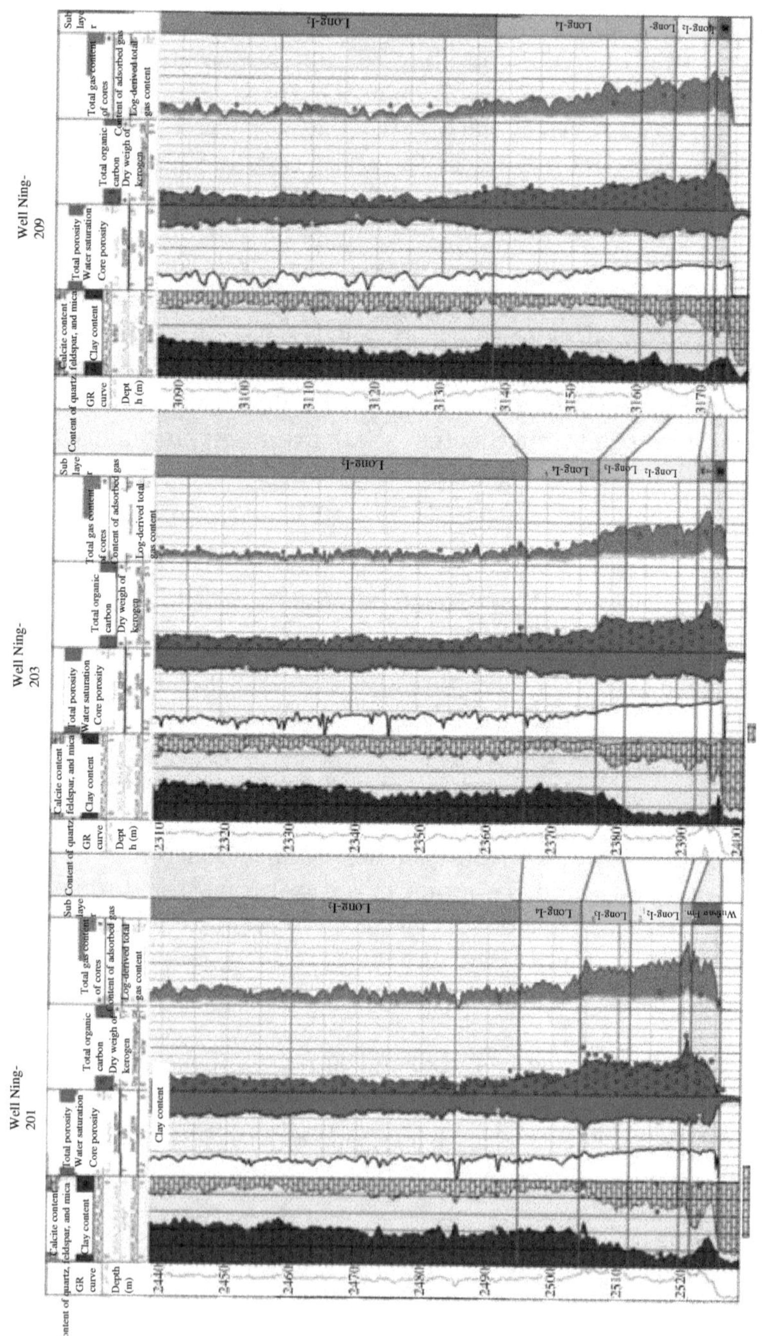

**Figure 7.6.** Integrated log evaluation map of wells Ning-201, Ning-203, and Ning-209.

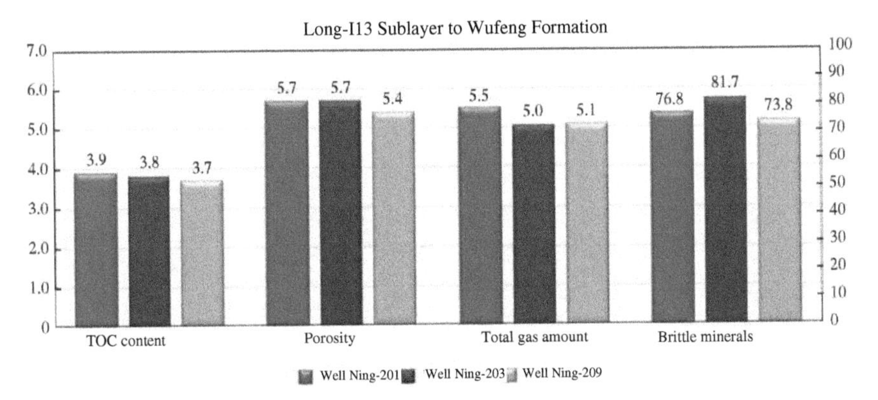


**Figure 7.7.** Histogram of parameters related to reservoir quality of wells Ning-201, Ning-203, and Ning-209 derived from log evaluation.

(2) Conduct the horizon interpretation of a long well tie seismic section and establish the structural interpretation framework of the main horizons;
(3) Conduct manual interpretation of 8 in lines ×8 cross lines based on the backbone section to establish the framework constraints for automatic horizon tracking through human–computer interactions;
(4) Achieve the horizon interpretation results of 1 in line ×1 cross line; then inspect the arbitrary seismic sections to ensure the reliability of the horizon interpretation.

The process of fault interpretation is similar to that of horizon interpretation. That is, fault interpretation is conducted from the backbone section to the framework section, with the scale becoming increasingly detailed, fine, and clear. Moreover, quality control is conducted according to the horizon interpretation results and discontinuity detection results during the interpretation.

For fracture prediction, the azimuthal fracture prediction method using pre-stack seismic data based on the horizontal transverse isotropy (HTI) theory may not be applicable to vertically transversely isotropic (VTI) shale layers. Moreover, the conventional discontinuity detection based on post-stack seismic data cannot reveal the detailed regularity of possible fracture zones and micro-faults apart from faults.

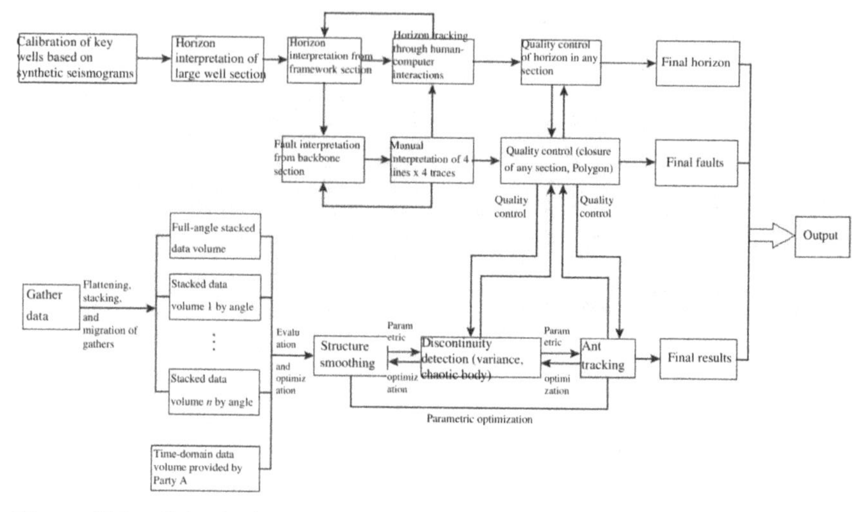


**Figure 7.8.** Seismic data interpretation and the technical route of ant tracking.

The ant-tracking technique is used to predict possible fracture zones and micro-faults in a block (Figure 7.8). This technique was rated as the best by the Best Exploration Technology Awards in 2005 by the journal *World Petroleum*. It is available in the attribute computation module of the Petrel software developed by Schlumberger, and the core technology of this module is the advanced ant-tracking algorithm. Ant colony members use pheromones to convey information between their nest and the food to find the most efficient path. Similarly, the ant-tracking algorithm populates a seismic volume with computer agents to let them move forward along the possible fault planes while sending out "pheromones" to mark the obvious fault planes but leave no or insignificant marks on the unlikely fault planes. The ant-tracking technique has been widely used in the automatic identification of faults based on post-stack seismic inversion. It was applied to the identification of potential fracture zones in shale reservoirs in the Huangjinba block, achieving good application performance.

The typical process of the ant-tracking technique includes four steps. First, highlight the boundary characteristics through the preprocessing of seismic data. Second, further enhance the boundary characteristics through edge detection. Third, establish the ant-tracking attributes by optimizing different parameters of ant tracking. Finally, obtain the final anomaly data volumes by verifying the reliability of the ant-tracking attributes.

To obtain reliable ant-tracking results, a prerequisite is to use seismic data volumes with a high signal/noise ratio as input. In this study, the pre-stack gathers were preprocessed through denoising, flattening, and amplitude equalization, and then the gathers with angles of 10°–28° with high signal/noise ratio and slight varying amplitude were stacked as the input of ant tracking.

Figure 7.9 is a map showing the ant-tracking attribute results along the bottom horizon of the Upper Ordovician. This figure shows that there are significant three fault systems with different strikes, which agree well with the manual interpretation results, thus verifying the accuracy of the manual interpretation.

Figure 7.10 shows the comparison between the microseismic event points and the ant-tracking results of wellpad H3. According to this figure, the locations of natural fractures are indicated by the microseismic events with magnitude greater than 1 agree well with the positions of anomalies identified by ant tracking, especially at the positions indicated by ① ② ③ ④ in the figure. Furthermore,

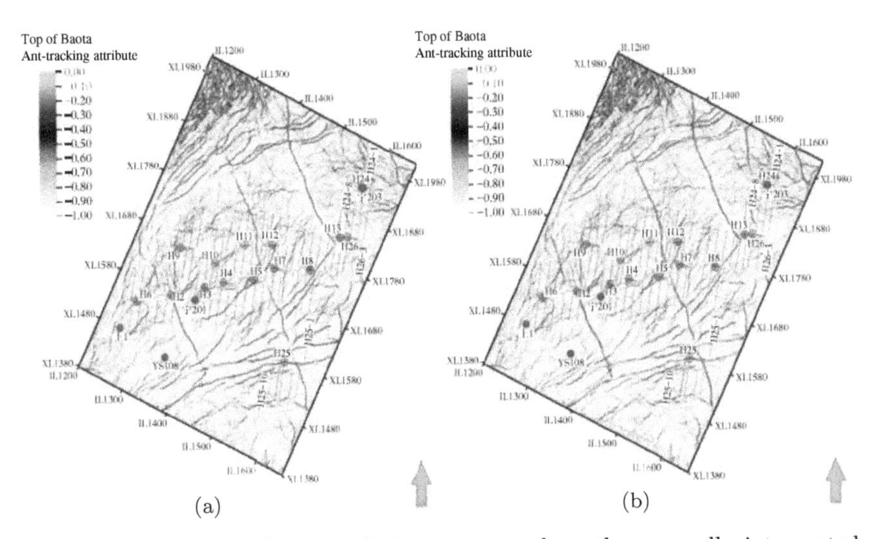


**Figure 7.9.** Ant-tracking attributes superposed on the manually interpreted faults on the bottom horizon of the Upper Ordovician in the Ning-201 well area. (a) Faults interpreted from ant tracking. (b) Manually interpreted faults.

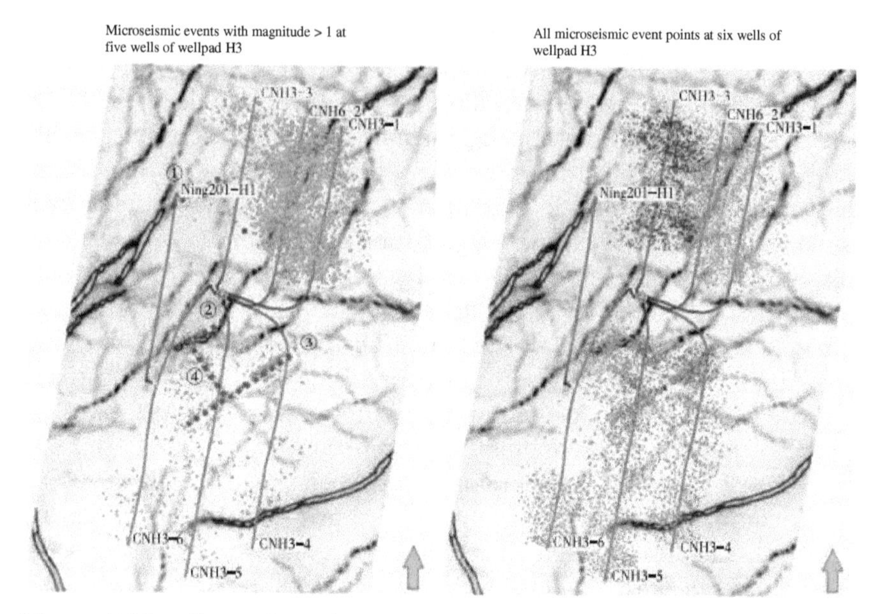


**Figure 7.10.** Comparison of microseismic event points and ant-tracking results of wellpad H3.

the hydraulic fracturing operations of well Ning-201-H1 indicated the presence of natural fractures at position ①.

### 7.2.4.3. *Pre-stack seismic inversion and calculation of key evaluation parameters*

Seismic inversion is critical in the geophysical study in the integrated study process. It takes the seismic, horizon, and log data as the basic input and outputs the elastic parameters of reservoirs after the testing and analysis using various inversion methods. Then, it further calculates reservoir parameters such as lithology, porosity, and TOC content based on the petrophysical analysis, thus providing the 3D trend constraints for the precise geological modeling and geomechanical study.

Considering the data basis and the study goal, this study, targeting the high-quality shale intervals of the Longmaxi–Wufeng formations and using AVO characteristics of the seismic gathers, adopted the pre-stack synchronous AVO inversion method to carry out log quality control and petrophysical analysis, the quality control and preprocessing of seismic data, fine-scale log-seismic calibration,

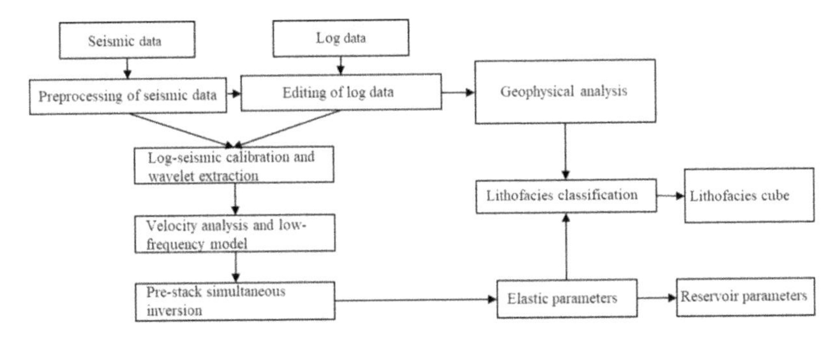


**Figure 7.11.** Flow chart of seismic inversion.

wavelet extraction, velocity field analysis, the generation of a low-frequency model, and the testing of various inversion methods, calculated the P-wave impedance and the $v_p/v_s$ ratio, and used them to calculate the reservoir parameters.

The process of seismic inversion is shown in Figure 7.11. With the log, seismic, and horizon interpretation data as input, elastic parameters (e.g., P-wave impedance, $v_p/v_s$ ratio, and Poisson's ratio) were obtained using the pre-stack synchronous inversion after quality control and analysis. Then, the distribution characteristics of physical property parameters of high-quality shales (e.g., thickness distribution, porosity, and TOC content) were further calculated using the lithological classification method. The study steps are discussed in detail in what follows.

(1) **Petrophysical analysis:** Petrophysical analysis is used to establish the relationships between subsurface rock attributes such as porosity, saturation, and shale content with the seismic response and to generate a theory to predict the physical properties of the rocks based on seismic inversion data. Specifically, standardarized analysis and correction of the petrophysical and geophysical log curves are carried out to ensure the consistency of log curves, thus providing accurate input data for wavelet estimation and seismic inversion. Meanwhile, the petrophysical analysis of the acoustic response of the target reservoir can be used to determine whether high-quality gas-bearing shales can be distinguished from tight shales and marls using petrophysical properties and seismic inversion. Afterward, the TOC and total gas contents of reservoirs can be predicted.

The log curves critical to seismic inversion are those of P-wave velocity, S-wave velocity, and density. However, the density and sonic log data have shallow investigation depth and tend to be affected by irregular wellbores and the invasion of drilling-fluid filtrate. Since the log data were well controlled and processed at the log interpretation stage in this study, the petrophysical analysis was only used to process the anomalies caused by end effects and imaged wellbores. Afterward, numerical methods (e.g., neural network method, multiple linear regression method, and general empirical relation method) were used to fit the log data and correct the abnormal parts of the curves, ensuring that all curves were reliably preprocessed before log-seismic calibration and wavelet extraction.

Figure 7.12 shows the corrected cross plots of the P-wave impedance and the $v_p/v_s$ ratio of the Longmaxi–Wufeng formations at the locations of two wells, with effective porosity and TOC content as color scales, respectively. On the cross plots, high-quality shales correspond to low P-wave impedance and low $v_p/v_s$ ratio; tight shales correspond to medium P-wave impedance and medium $v_p/v_s$ ratio, while marls are associated with high P-wave impedance and high $v_p/v_s$ ratio. Therefore, it can be inferred that high-quality shales can be distinguished from tight shales and marls to a certain extent using the P-wave impedance and the $v_p/v_s$ ratio obtained from seismic inversion.

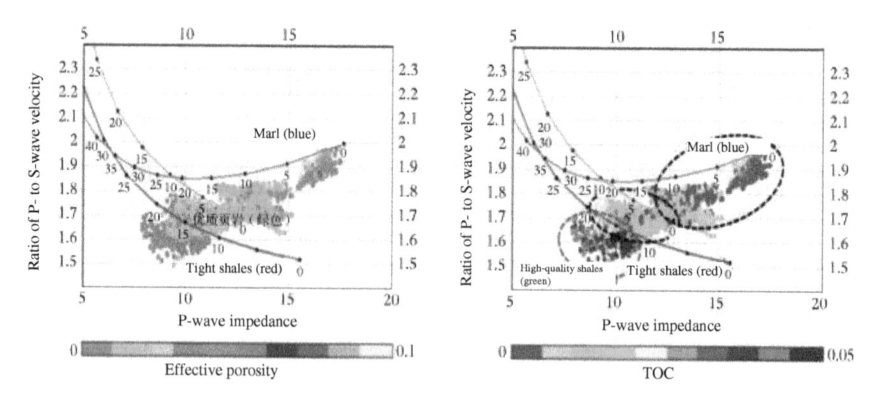


**Figure 7.12.** Cross plot of P-wave impedance and ratio of P-wave to S-wave velocity for the target intervals at two wells.

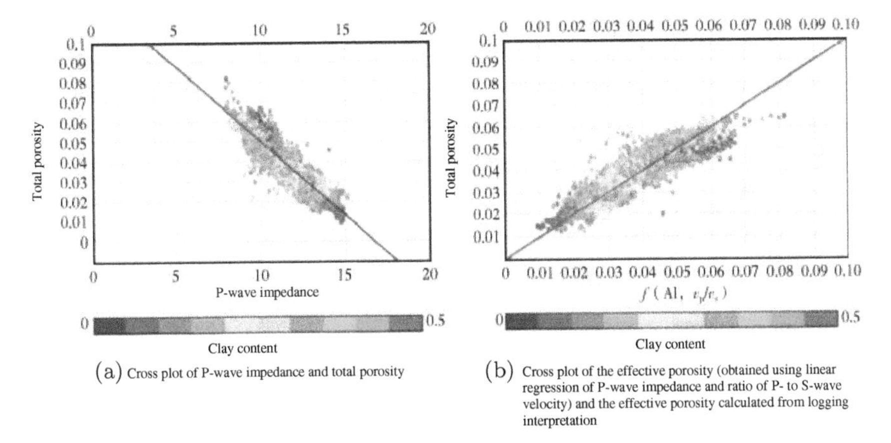


(a) Cross plot of P-wave impedance and total porosity

(b) Cross plot of the effective porosity (obtained using linear regression of P-wave impedance and ratio of P- to S-wave velocity) and the effective porosity calculated from logging interpretation

**Figure 7.13.** Analysis of reservoir porosity.

Figure 7.13 shows the analysis results of the reservoir porosity of the target intervals at two key wells, including the cross plot of P-wave impedance and total porosity and the cross plot of effective porosity and the effective porosity. Among them, the predicted effective porosity was obtained from the linear regression of P-wave impedance and $v_p/v_s$ ratio, while the measured effective porosity was determined by logging interpretation. Both cross plots use shale content as the color scale. According to this figure, there is a significant negative correlation between the total porosity and the P-wave impedance, with a correlation coefficient of up to $-0.9$. Moreover, there is a weak correlation between the effective porosity and P-wave impedance or between the effective porosity and the $v_p/v_s$ ratio. However, the correlation coefficient between the effective porosity predicted using the P-wave impedance and the $v_p/v_s$ ratio and the measured porosity is up to 0.85. Therefore, the 3D porosity can be accurately predicted based on the linear regression and the seismic inversion results.

The petrophysical analysis reveals that there is the strongest correlation between the TOC content and the density of the shale gas at the target intervals of the Changning block, with a correlation coefficient of up to $-0.96$. However, the density results from seismic inversion are often inferior to inverted P-wave impedance and $v_p/v_s$ ratio. Therefore, the linear regression method of P-wave impedance and the $v_p/v_s$ ratio was adopted to predict the

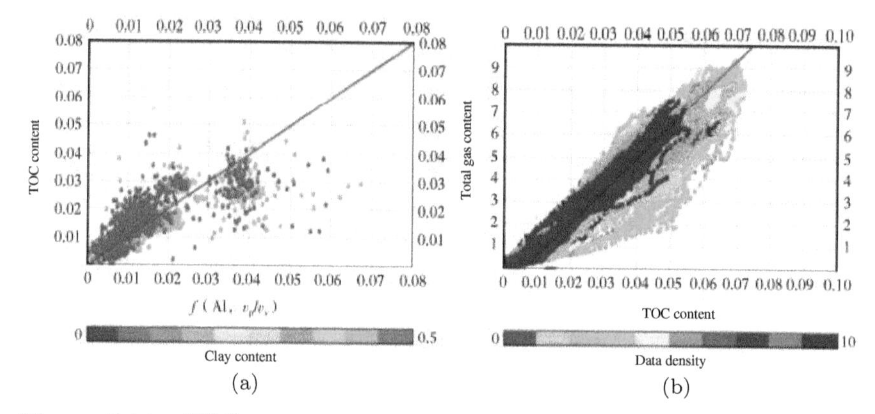


**Figure 7.14.** TOC content prediction method. (a) Cross plot of measured and predicted TOC content. (b) Cross plot of TOC content and total gas content.

TOC content. Figure 7.14(a) shows the cross plot of measured TOC content and predicted TOC content, which have correlation coefficient of 0.83. Figure 7.14(b) shows the cross plot of TOC content and total gas content for the target intervals at 15 wells. In general, high TOC content corresponds to high gas content and high porosity, and the linear correlation coefficient between the TOC content and total gas content reaches 0.96. Therefore, the 3D TOC and total gas contents can be predicted based on the linear regression and the seismic inversion results, thus providing a data basis for subsequent geological models, geomechanical models, and comprehensive analysis.

(2) **Quality control of seismic data:** Seismic data are crucial to seismic inversion. During the inversion, extensive control quality and preprocess were conducted on the seismic data in detail, including noise suppression, flattening, stacking, and amplitude equalization of gather data, the division of angle stack scheme, and the alignment of angle stack data.

(3) **Key process of seismic inversion:** Seismic inversion is the most important means to obtain key reservoir parameters. Its key steps include wavelet estimation, the generation of a 3D low-frequency model, the testing and determination of the parameters for seismic inversion, and quality control of the AVO inversion results. The principle of seismic inversion is described in the preceding sections and thus is not discussed here.

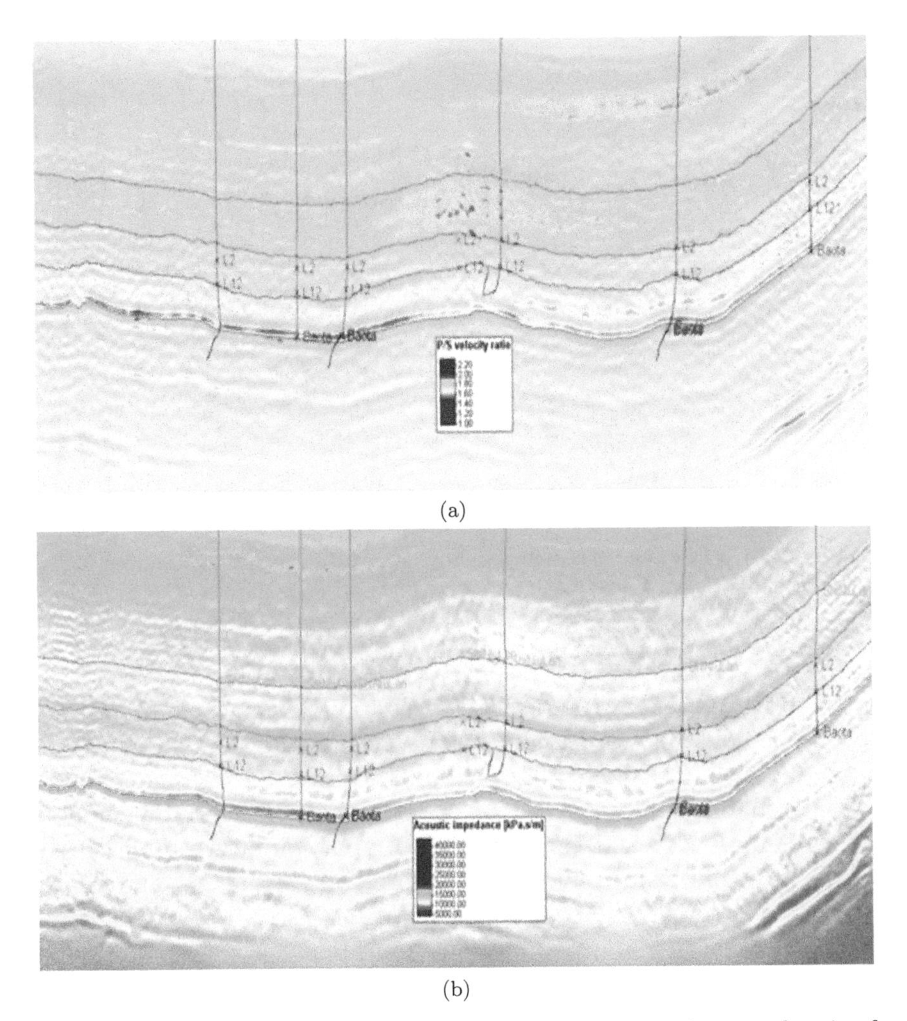


**Figure 7.15.** Well tie seismic sections of the P-wave impedance and ratio of P-wave to S-wave velocity. (a) Well tie seismic section of the P-wave impedance obtained from the pre-stack synchronous AVO inversion. (b) Well tie seismic section of the ratio of the P-wave to S-wave velocity obtained from seismic inversion.

Figures 7.15(a) and 7.15(b) show the cross well section of P-wave impedance obtained from the pre-stack synchronous AVO inversion and the cross well section of the inverted $v_p/v_s$ ratio obtained, respectively. According to these figures, the Long-I$_1$ Submember consists of a set of reservoirs with relatively low

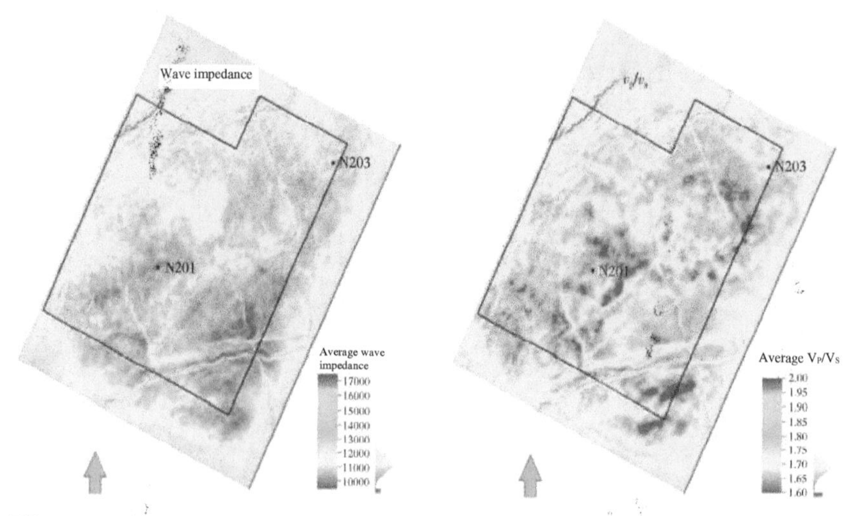


**Figure 7.16.** Plans of P-wave impedance and ratio of P-wave to S-wave velocity of the Long-$I_1$ Submember.

P-wave impedance and a low $v_p/v_s$ ratio, which are relatively continuous but vary to a certain extent in the lateral direction.

Figure 7.16 shows the plans of the inverted average P-wave impedance and average $v_p/v_s$ ratio of the Long-$I_1$ Submember.

Figures 7.17 and 7.18 show the comparison between measured P-wave impedance and $v_p/v_s$ ratio of wells Ning-201 and Ning-203, respectively. The P-wave impedance is on the left and the $v_p/v_s$ ratio is on the right. The red curves represent the results of the measured P-wave impedance and $v_p/v_s$ filtered to the seismic frequency, and the blue curves denote the inverted P-wave impedance and $v_p/v_s$ ratio. It can be concluded that the inverted and measured data of the two wells are highly consistent with each other at the target interval — the Longmaxi Formation, indicating that the inversion results are highly reliable.

(4) **Calculation of reservoir parameters:** According to the above description, the petrophysical analyses of various lithologies were performed based on the log data, and the 3D seismic elastic parameters such as P-wave impedance and the $v_p/v_s$ were obtained using seismic data. Converting the 3D elastic parameters obtained through seismic inversion into the physical

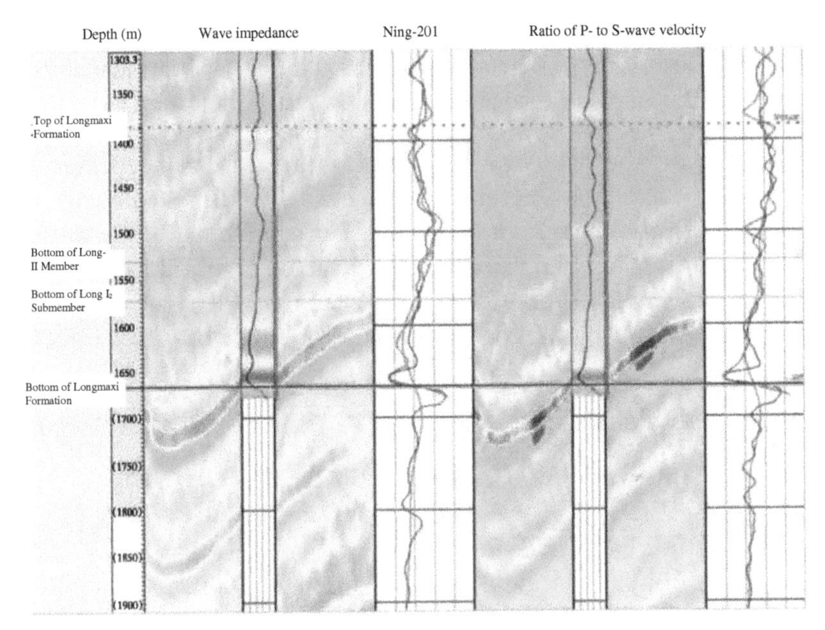


**Figure 7.17.** Wave impedance and ratio of P-wave to S-wave velocity of the near-well seismic traces of well Ning-201.

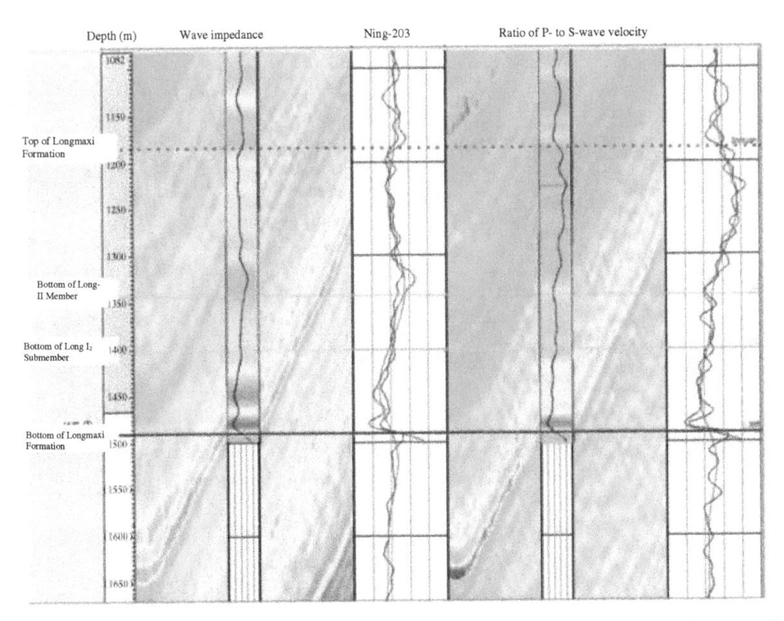


**Figure 7.18.** Wave impedance and ratio of P-wave to S-wave velocity of the near-well seismic traces of the well Ning-203.

property parameters of the reservoirs based on the elastic characteristics of various lithologies is an important step in reservoir prediction using the seismic inversion method.

The LithoCube lithology prediction is a method to convert the inverted 3D elastic parameters into the corresponding lithologies and lithological probability volumes using the probability density functions of elastic parameters of various lithologies obtained from log data, as shown in Figure 7.19. This study divided the lithologies into three types, according to calcareous content, porosity, and TOC content.

High-quality shales: $V_{\mathrm{carb}} < 0.5$, TOC content $\geq 0.02$, $\emptyset \geq 0.02$.

Tight shales: $V_{\mathrm{carb}} < 0.5$, TOC content $< 0.02$, $\emptyset < 0.02$.

Marls: $V_{\mathrm{carb}} \geq 0.5$.

$\lambda\rho$, $\mu\rho$, and Poisson's ratio were determined as the optimal parameter combination through the optimization of sensitive attribute parameters. The prediction success rates of high-quality shales (red), tight shales (green), and marls (blue) were 89.4%, 68.5%, and 79.7%,

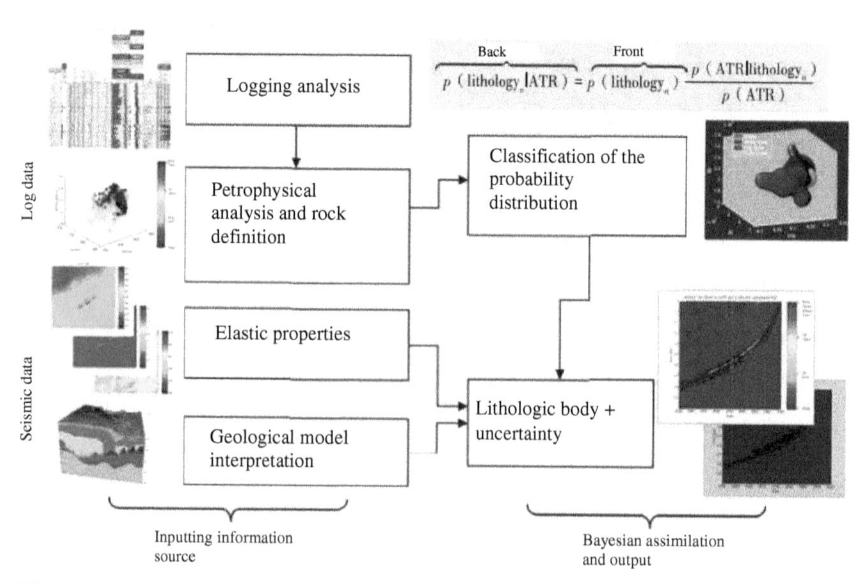


**Figure 7.19.** Flow chart of lithology prediction process using the LithoCube method.

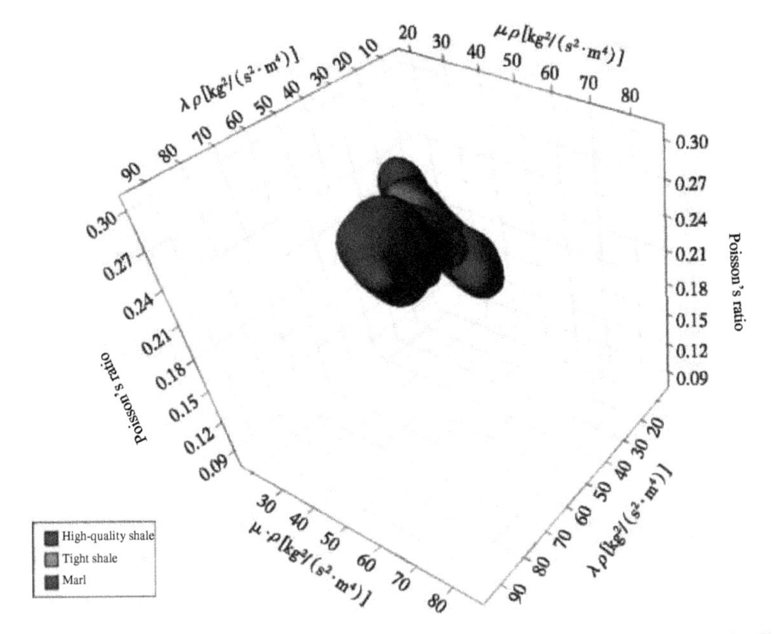


**Figure 7.20.** 3D probability density function obtained using the LithoCube method.

respectively. The 3D probability density functions are shown in Figure 7.20.

Figures 7.21 and 7.22 show the lithology prediction results and the probability volume of high-quality shales, respectively. According to these figures, high-quality shales are mainly distributed in the Long-$I_1$ Submember, and this finding is consistent with the current understanding obtained from drilling. It is noteworthy that, due to the limitations of the seismic resolution, only the average characteristics of the Long-$I_1$ Submember can be identified from the results of the seismic inversion. The reservoir changes on a smaller scale will be identified and analyzed in the later fine-scale geological modeling.

Figure 7.23(a) shows the thickness of high-quality shales obtained based on probability. Figure 7.23(b) shows a scatter diagram of the annual cumulative gas production of 12 horizontal wells and the average thickness of high-quality shales extracted along the trajectory of the target intervals at the horizontal wells, indicating a close correlation between the two parameters.

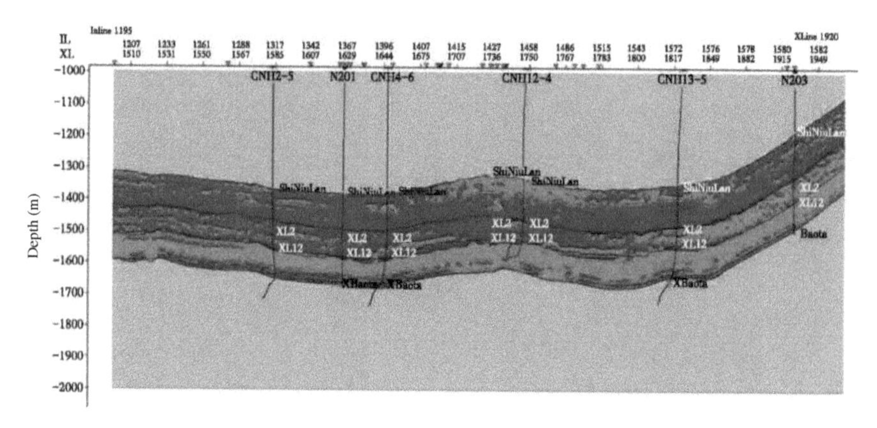


**Figure 7.21.** Well tie seismic section of lithologies.

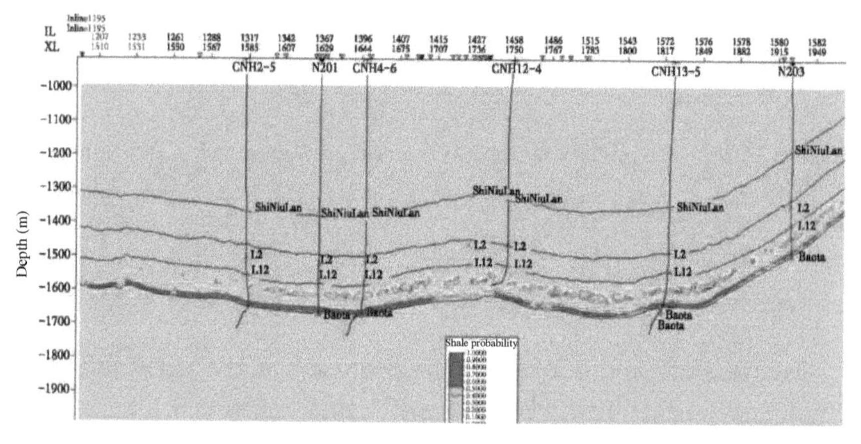


**Figure 7.22.** Well tie seismic section of the probability of high-quality shales.

The calculation formulas for porosity and TOC content were derived from log curve fitting in the aforementioned petrophysical analysis. Many fitting methods and stratum-based statistics calculations were tried in this study, and the formula yielding the highest correlation coefficient was finally selected. The porosity was fitted using the P-wave impedance, with a correlation coefficient of −0.9. The TOC content was fitted based on both the P-wave impedance and the $v_p/v_s$ ratio, with a correlation coefficient of 0.83. The well tie seismic sections of porosity and the TOC content are shown in Figures 7.24 and 7.25, respectively.

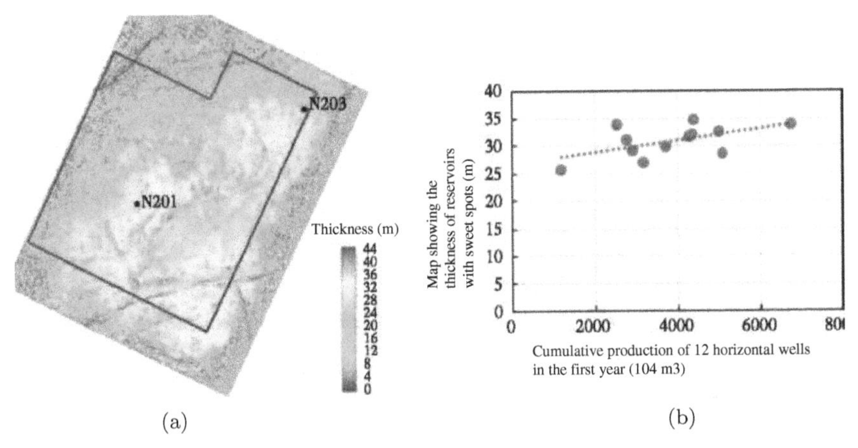


**Figure 7.23.** Comparison of the annual cumulative gas production and the average penetrated thickness of high-quality shales at the locations of 12 horizontal wells. (a) High quality shale thickness map. (b) Scatter plot of average thickness of high-quality shale.

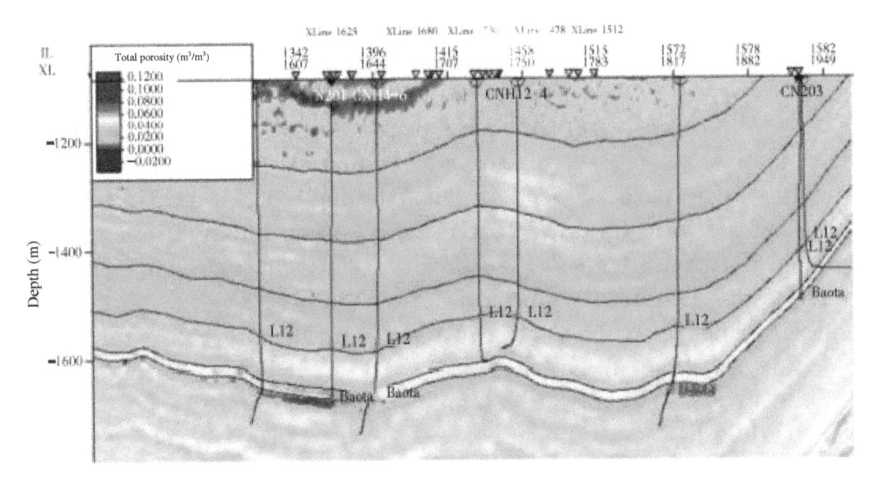


**Figure 7.24.** Well tie seismic section of porosity.

Since the preceding petrophysical analysis shows that there is a significant linear relationship between the total gas content and the TOC content, this correlation can be used to further convert the TOC content volume into the total gas content volume. However, uncertainty was introduced in each conversion process and was cumulated and amplified. Therefore, the converted total gas content

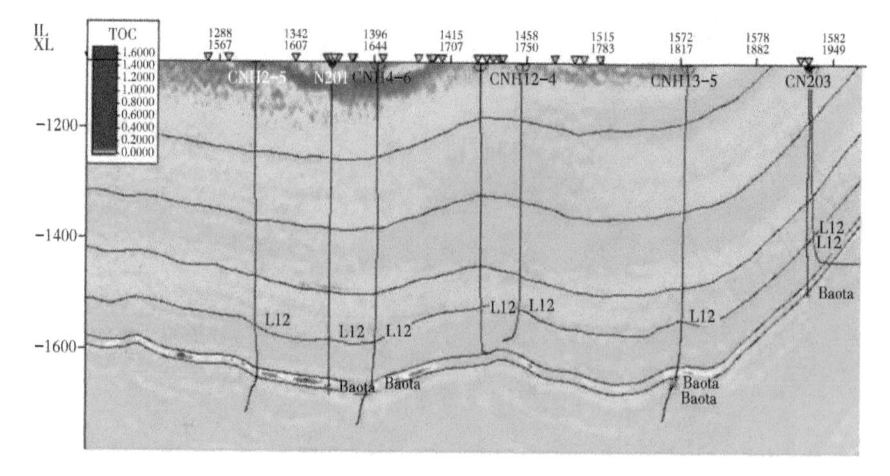


**Figure 7.25.** Well tie seismic section of TOC content.

volume can be only used to qualitatively analyze the variation trend thorough the study area. The well tie seismic sections of total gas content and Young's modulus are shown in Figures 7.26 and 7.27, respectively.

Overall, after several rounds of inversion and testing and parameter optimization, the P-wave impedance and $v_p/v_s$ ratio at the target intervals were highly consistent with the log data, indicating highly reliable 3D elastic volumes. Moreover, based on the inversion results, reliable 3D parameters (e.g., porosity, TOC content, and gas content) were calculated using correlations between these parameters and the impedance and $v_p/v_s$ ratio of the target intervals established by comprehensive analyses.

The inversion results that can be used for quantitative evaluation include P-wave impedance, $v_p/v_s$ ratio, porosity, TOC content, gas content, and lithology probability volumes. The inversion results that can be used for qualitative evaluation include volumes of density, lithology, and brittleness index. The inverted reservoir parameters, such as porosity and TOC content, can support the selection of favorable areas and well location design from the perspective of reservoir quality. Moreover, the inverted 3D elastic parameters can be used for further fine-scale geological modeling and geomechanics-related studies.

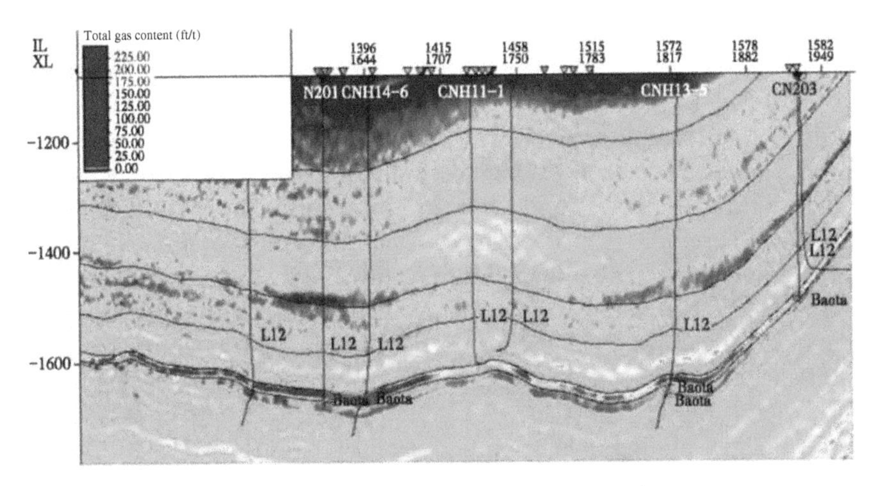


Figure 7.26. Well tie seismic section of the total gas content.

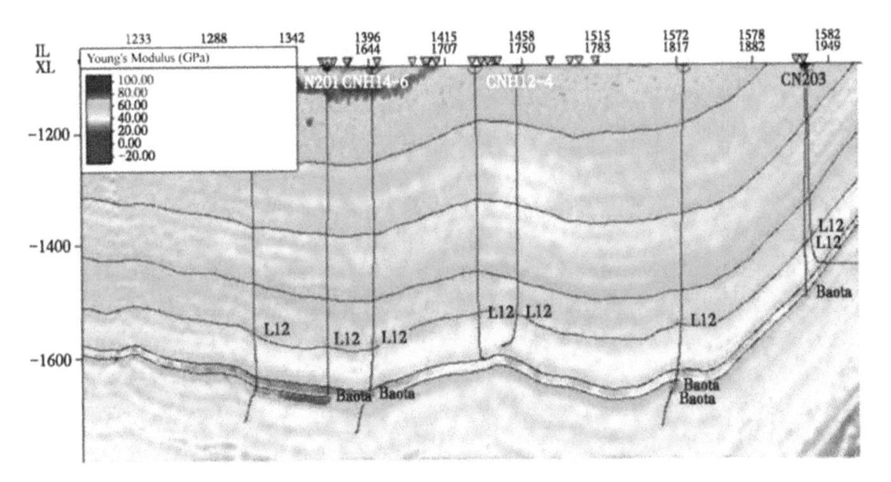


Figure 7.27. Well tie seismic section of Young's modulus.

### 7.2.4.4. *Geological modeling of reservoirs*

The area of well Ning-201 has a complex geology and, thus, it is necessary to conduct the 3D quantitative characterization of reservoirs for the efficient development of shale gas fields in this area. For horizontal wells, it is difficult to achieve accurate sublayer correlation using only a single gamma ray curve in determining the target horizons along their trajectories. They are different from vertical wells in this regard. This poses a challenge for the description of structural reconstruction and the calculation of penetration probability. If the geological horizons along the well trajectories cannot be determined, the geosteering and fracturing operations will lose definite targets. Therefore, it is necessary to conduct 3D geological modeling to describe reservoirs in detail. The present abundant seismic, log, and core analysis data as well as the drilling and completions of nearly 70 wells provide a good data basis for fine-scale reservoir characterization and 3D geological modeling of the area of well Ning-201 (Figure 7.28).

Geological modeling is the integration of various geological understandings and serves as an important intermediate link. The research philosophy of geological modeling is as follows: (1) integrating single-well data, test data, and seismic data to create a structural model through fine-scale correlations and seismic interpretation; (2) building reservoir attribute models with log data interpretation and seismic inversion, and (3) establishing a fracture model by combining seismic attributes and image logs.

(1) **Fine-scale sublayer correlation:** A fine-scale sublayer correlation was carried out for 49 horizontal wells in the area of well Ning-201, obtaining the log data of 41 horizontal wells collected on completion of drilling and the logging while drilling (LWD) data of eight wells. Based on the log data collected on completion of drilling as well as the combined gamma ray, resistivity, density, sonic, and neutron curves, various sublayers can be effectively determined.

Therefore, this study proposed an integrated sublayer correlation method, aiming to conduct the sublayer correlation and establish a geosteering model by integrating log and seismic data as well as multidimensional data (e.g., 2D, 3D, and TST-domain data). The main steps are as follows.

(i) The comparison of cycles based on the (TST) domain. For horizontal wells, it is difficult to conduct sublayer correlation on sections of a length range along the dip direction or true vertical depth since these wells are affected by variations in well inclination and the dip angles of strata. Therefore, the

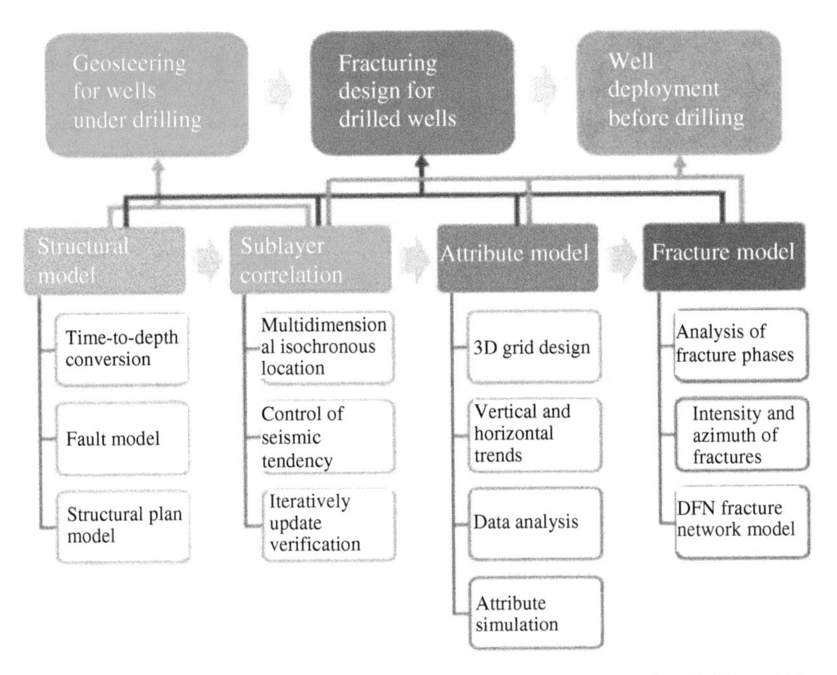


**Figure 7.28.** Geological modeling flowchart for the area of well Ning-201.

stratigraphic locations of the trajectories of horizontal wells can be determined only by estimating the true thickness of strata based on the structural dip angles in seismic data and the variations in cycles of GR curves. In this way, the output results include the data on the sublayers and dip angles of individual wells. For instance, the trajectory of well H9-1 can be subdivided into downcut, upcut, and downcut sections. After the true thicknesses of the three sections were determined, the characteristics of the cycles of the GR curve of this well can be compared with those of vertical wells, as shown in Figure 7.29. After the TST domain calculations, the stratigraphic locations of well trajectories can be determined more accurately, thus providing an accurate and intuitive reference for fracturing design after drilling.

(ii) The dip angles estimated in the TST domain-based sublayer correlation can be used to build a 2D geosteering section. A simulated GR curve can be calculated according to the structural section in the geosteering forward model. Then the dip angles can be calculated by combining the simulated

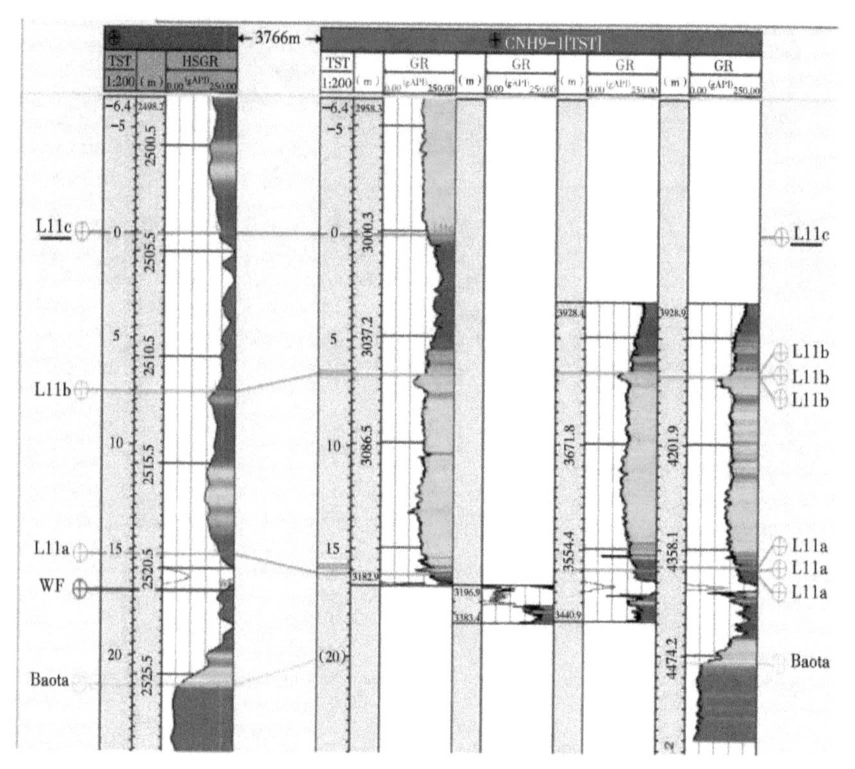


**Figure 7.29.** TST domain-based sublayer correlation between wells Ning-201 and H9-1.

GR curves with the measured GR curves. The resulting 2D geosteering section is also the basis for the 3D structural modeling (Figure 7.30).

(iii) By comparison with the 2D geosteering section, the seismic section can present subtle structural trends and can well reflect microstructural and fault characteristics. Taking well H4-5 as an example, the dip angles of strata at its landing point are 4°–7° (downdip), and its trajectory remains at the Long-$I_1^3$ Sublayer near the landing point. At a depth greater than 3,700 m, the structures along the well trajectory become steep until a maximum dip angle (downdip) of 14°. Afterward, the strata become gentle, with a dip angle of 5°–7°, and the well trajectory accordingly enters the lower Wufeng Formation — Long-$I_1^2$ Sublayer, exhibiting significant low-amplitude folds on the geosteering section. Such

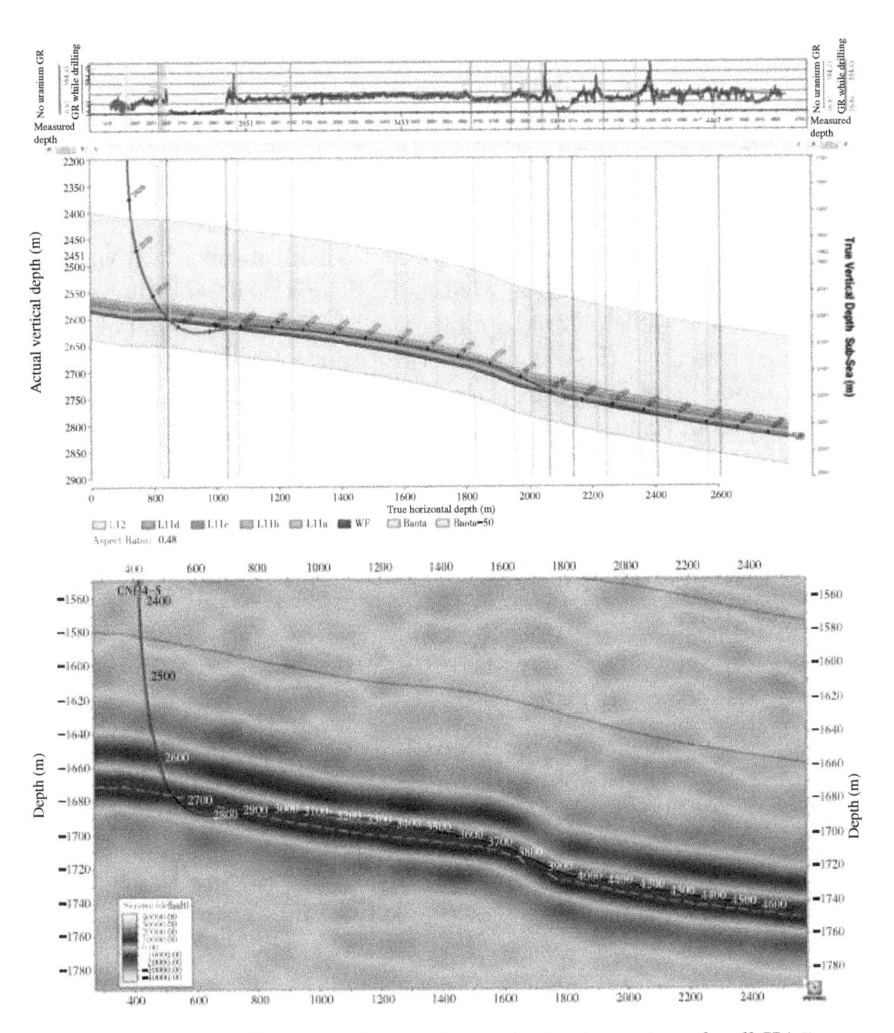


**Figure 7.30.** 2D geosteering section and seismic section of well H4-5.

structural characteristics are also reflected in the seismic section, indicating that the seismic data can well reflect microstructural characteristics.

(iv) The verification from adjacent wells helps overcome the uncertainty of the time-to-depth relationship in the seismic data of mountainous areas, thus continuously improving the accuracy of the structural model. Moreover, multidimensional interactions, as well as various comparison and display methods such as 2D, 3D, and TST domains, also provide different means for sublayer correlations and geosteering while drilling.

(2) **Structural modeling:** The time-to-depth conversion is the basis of structural modeling. The time-to-depth conversion method used in this stage is to build and correct the velocity model based on the time-to-depth calibration of 10 vertical wells in the study area as well as 70 drilled wells and more than 200 virtual wells arranged along the horizontal portions of these wells. Based on the seismic interpretation of horizons, the velocity model, starting from the seismic reference datum, covers the Hanjiadian, Shiniulan, Longmaxi, Wufeng, and Baota Formations. The depth-based structural model includes 36 faults and boundaries of eight sublayers. In this model, the bottom of the Wufeng Formation, the top of the Long-I Member, and the top of the Long-II Member were controlled by the stratigraphic boundaries obtained from the interpretation of the seismic data, while the sublayers of the Long-$I_1$Submember were determined based on their thicknesses and plans.

Following the time-to-depth conversion, faults were modeled using the Pillar Gridding method by correctly reproducing relationships between faults. In the structural modeling, the results of the 2D geosteering section were applied to ensure an accurate match between the structures along the well trajectories, especially those in the horizontal portions of well trajectories. The structural model is the result of the modeling based on the log-seismic integration, as well as the application of the seismic interpretation of faults, stratigraphic boundaries, single-well stratigraphic divisions, and geosteering sections.

For the grids of the structural model, the planar grid cell size was set at 30 m × 30 m, which can be increased to 60 m × 60 m for the subsequent simulation model if required. Moreover, the vertical cell size varied considering the logs' vertical resolution, the shales' vertical heterogeneity, and the thickness of the strata such as the Guanyinqiao Member (approximately 0.5 m). Specifically, the average grid cell thickness was 0.5 $m^2$ for the Long-$I_1$ Submember and the Wufeng Formation and gradually increased to 5 m upward, yielding 63.87 million grid cells with a size of 30 m × 30 m × 0.5 m in total. Since the bottom of the Baota Formation was not drilled, the interval below 50 m down from the top of the Baota Formation was considered the limestone interval of the Baota Formation in the model. The proportional vertical gridding of the model was conducted according to the isochronous principle of sequence stratigraphy.

The results of the structural modeling mainly include the map showing the burial depth of the bottom of the Wufeng Formation, the structural maps of various sublayers, the map showing the dip angles of strata, and the structural curvature map. Figures 7.31 and 7.32 are the structural map of the bottom of the Wufeng Formation and its structural dip angel map. The combination of the two maps can provide an intuitive distribution of the structural characteristics and complexity for well deployment and drilling design.

(3) **Attribute modeling:** A 3D reservoir quality attribute model is designed to evaluate the 3D spatial distribution of reservoirs to support horizontal wells' deployment and fracturing design.

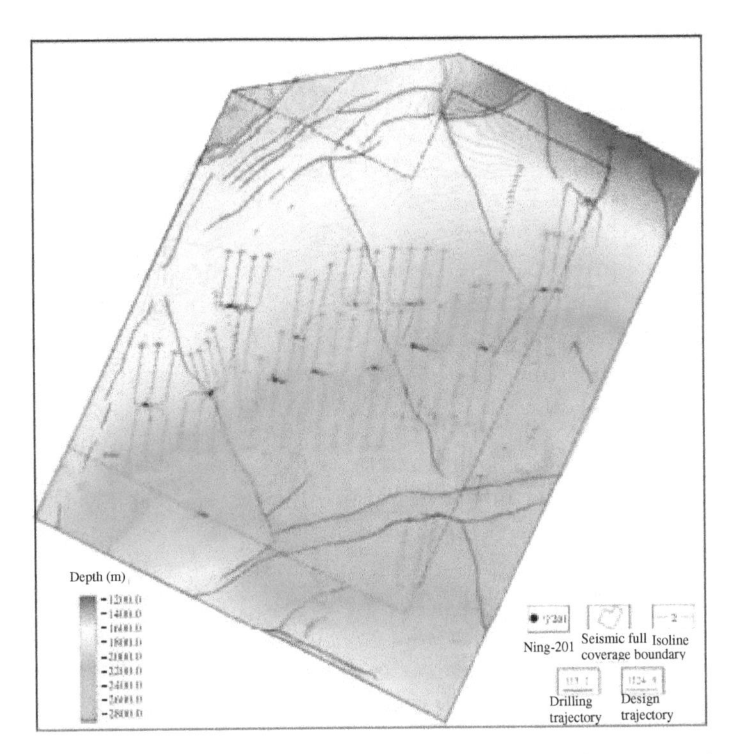


**Figure 7.31.** Structural map of the bottom of the Wufeng Formation.

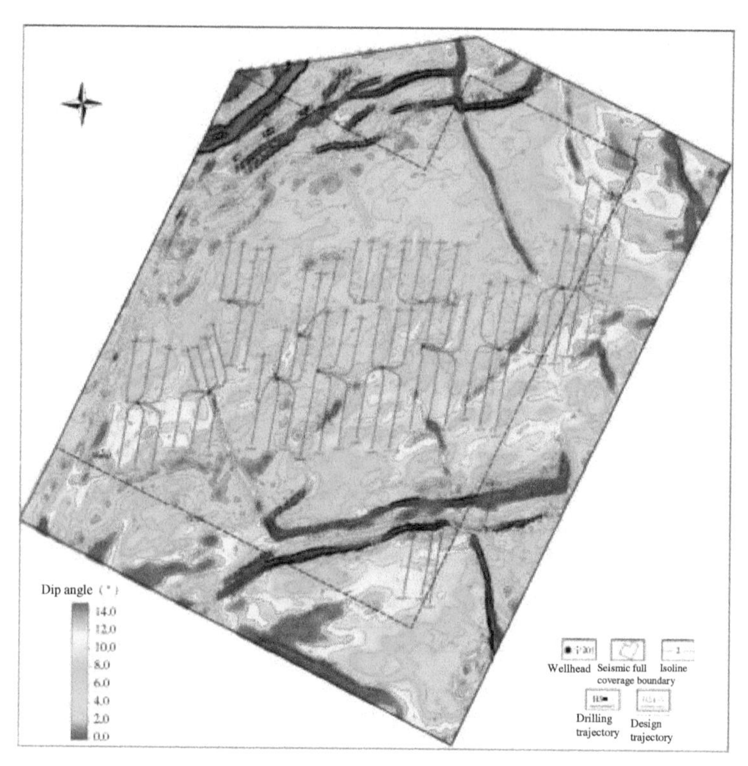


**Figure 7.32.** Structural dip angle map of the bottom of the Wufeng Formation.

The attribute modeling method used in the study area was to create 3D attribute volumes based on seismic inversion and single-well log interpretation to serve the 3D geomechanical simulation and fracturing engineering. The workflow is shown in Figure 7.33. A 3D attribute model can be built in four steps:

(i) *3D grid design*: The horizontal size of grid cells is determined by combining seismic graphic primitives, and their vertical size is determined based on log resolution.
(ii) *Upscaling of log curves*: The log curves are re-sampled to the grid cells through which the well trajectories pass;
(iii) *Resampling of the inverted attributes*: The inverted attribute volumes are resampled to the 3D grid cells;
(iv) *Attribute modeling based on the log-seismic integration*: The attribute models are built by using the inverted attributes as

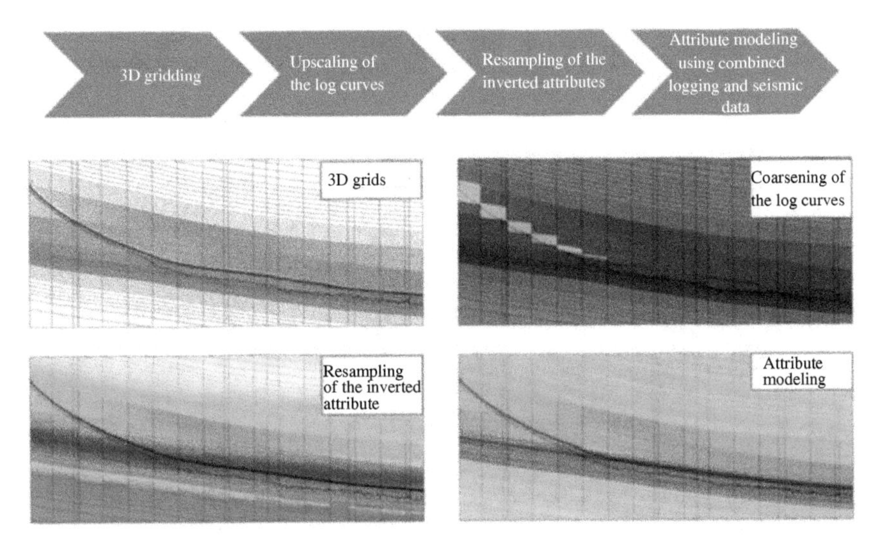


**Figure 7.33.** Technical flow chart.

soft data controlling the horizontal distribution of attributes and using log data as hard data controlling the vertical distribution of attributes.

Using the attribute modeling technology based on the seismic–geological–engineering integration, this study built models of the key evaluation parameters of shale gas reservoirs, including reservoir quality attributes (porosity, TOC content, gas saturation, clay content, gas content, and density), rock elasticity and strength parameters (transversal and vertical Young's modulus), Poisson's ratio (transversal and vertical), uniaxial compressive strength, tensile strength, internal friction angle, and brittleness index (Figures 7.34 and 7.35).

Based on the 3D attribute models, plans of the TOC content, porosity, gas content, clay content, and gas saturation of various sublayers can be prepared, providing an important basis for the subsequent design and adjustment of horizontal well trajectories and maximizing the probability of penetration of high-quality reservoirs.

Considering different data acquisition conditions and rock discontinuity scales, this study divided fractures and faults in the study area into three types (Figure 7.36).

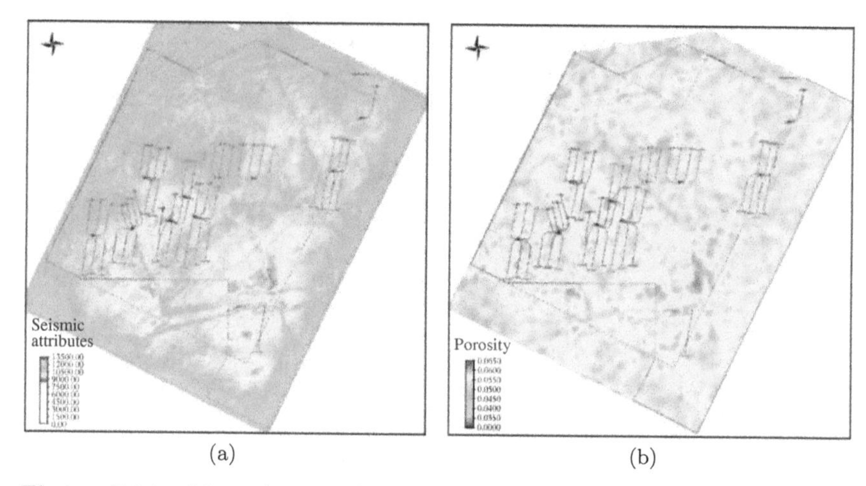


**Figure 7.34.** Maps showing the wave impedance inversion (a) and the total porosity of the model (b) of the Long-$I_1$ Submember — the Wufeng Formation.

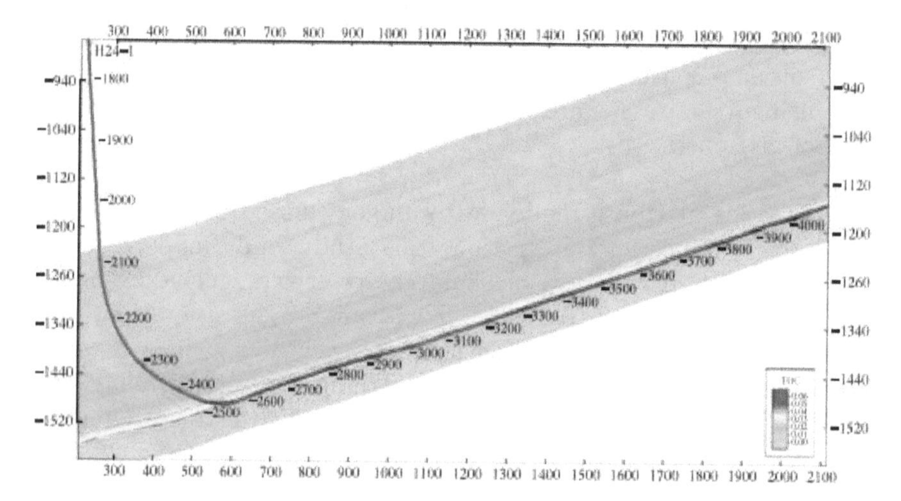


**Figure 7.35.** TOC content section along well H24-1.

Large-scale discontinuities: refer to several large faults in the study area. They are controlled by regional tectonic movements, extend for 5–10 km, and have strikes of NE-SW primarily and NW-SE secondarily. One or more dislocation events can be interpreted in the seismic section.

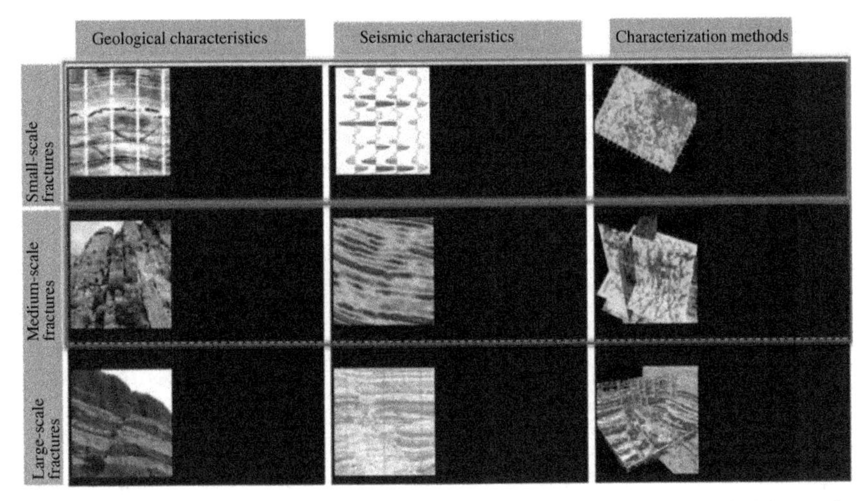


**Figure 7.36.** Characteristics of natural fractures and faults classified by scale.

Medium-scale fractures: 11 faults (or fracture zones) have been identified in the work area through ant tracking. It can be inferred that they are controlled by large faults from their strikes. They extend for about several hundred meters laterally. The events in the seismic section show changes in occurrence (distortion) or amplitude (weakening). Therefore, medium-scale fractures are shown in the seismic sections. However, they are present much clearer in the ant tracking results.

Small-scale discontinuities: refer to micro-fractures on a meter scale. They are controlled by faults and small folds and are visible in image logs and cores. They are difficult to be distinguished using seismic data but can be simulated by fracture modeling.

Large-scale faults are generally avoided in well deployment, while medium-scale faults or fracture zones have important effects on shale gas development. First, they pose challenges to drilling and geosteering. If shear sliding occurs in fracture zones or small faults, the stability of borehole walls will be greatly affected, and the resulting microstructures will affect geosteering. Furthermore, medium-scale fractures will affect the efficiency and propagation of hydraulic fractures. As for the characterization method, ant tracking was selected as the main method after a detailed comparison and screening.

The reliability of the ant-tracking results depends on the cross-validation of multiple data. This study validated the rationality of the ant-tracking attributes using various methods and various data. The former includes the inspection of seismic sections, the comparison of seismic attributes, and the comparison of image logs, and the latter includes drilling data and microseismic data.

The key to the natural fracture modeling system of the study area is to correctly locate micro faults of different scales, natural fracture zones, and small-scale discrete natural fractures based on the log-seismic integration. The ultimate goal of natural fracture modeling is to study the interactions between natural fractures and induced fractures to support hydraulic fracturing operations.

Regarding discrete fracture networks (DFNs), randomly generated fracture groups were directly used to form fracture networks, which are used to characterize the fracture system. The parameters required for DFN characterization include the intensity, azimuth, dip angle, and extended length and height of fractures.

Based on the above understanding, this study developed a DFN model, aiming to consider the effects of natural fractures in strata in subsequent simulations of rock mechanics and reservoirs. In this way, the subsurface geology can be well described (Figure 7.37).

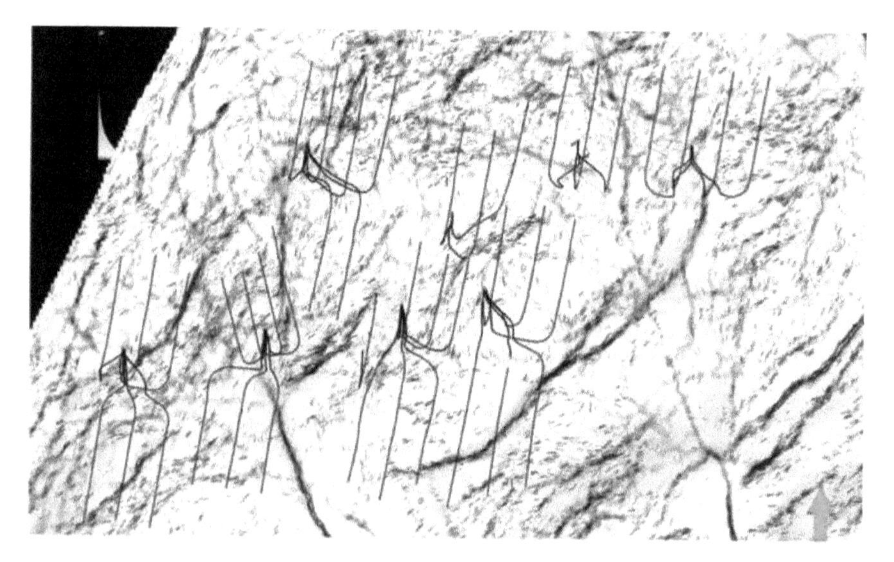

**Figure 7.37.** Superimposed map of DFN and ant-tracking attributes.

#### 7.2.4.5. *Geomechanical modeling*

A 1D geomechanical modeling was first conducted. An anisotropic TIV (vertically symmetric anisotropic medium) model was used to calculate the geomechanical parameters. Moreover, the anisotropic parameters were also considered in calculating the *in situ* stress using the pore elasticity formula. The input of the 1D geomechanical modeling mainly includes the log, test, and experimental analysis data. The process steps include inspecting input data; analyzing drilling complexity; calculating anisotropic parameters, elastic modulus, overburden pressure, pore pressure, rock strength, the maximum and minimum horizontal principal stresses and their directions; and analyzing borehole instability.

Based on the fine-scale 1D geomechanical modeling, the 3D seismic geomechanical modeling was carried out through seismic inversion, and the processes are as follows:

(1) A 3D finite element mesh was built according to the 3D geological model;
(2) The 3D geomechanical parameters, including Young's modulus, Poisson's ratio, uniaxial compressive strength, friction angle, and tensile strength, were determined based on seismic inversion results and 1D geomechanical modeling results.
(3) Faults and fracture zones were integrated into the model;
(4) The 3D pore pressure volumes were calculated;
(5) The original stress field, including minimum horizontal principal stress, maximum horizontal principal stress, and overburden pressure, was calculated using the 3D geomechanical software VISAGE developed by Schlumberger.

Additionally, to correctly simulate the boundary conditions of reservoirs, it is necessary to add overburden, underburden, and lateral rock layers outside the reservoirs. Figure 7.38 shows the 3D finite element mesh of the study area obtained according to the above processes. Figure 7.38(b) shows the model of the reservoirs in the study area, and Figure 7.38(a) shows the overall model containing overburden, underburden, and lateral rock layers.

Figures 7.39 and 7.40 show the comparison of the 1D and 3D geomechanical parameters and rock strength of wells Ning-201 and

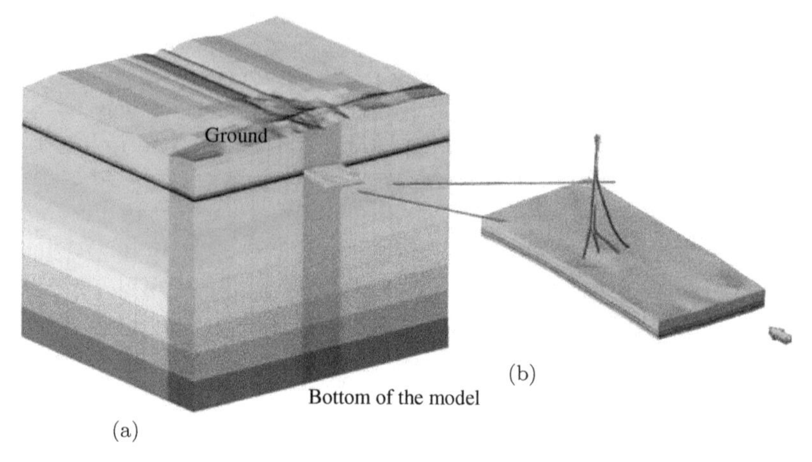


**Figure 7.38.** 3D finite element mesh of the H9 wellpad area. (a) Overall model. (b) Model of studied reservoirs.

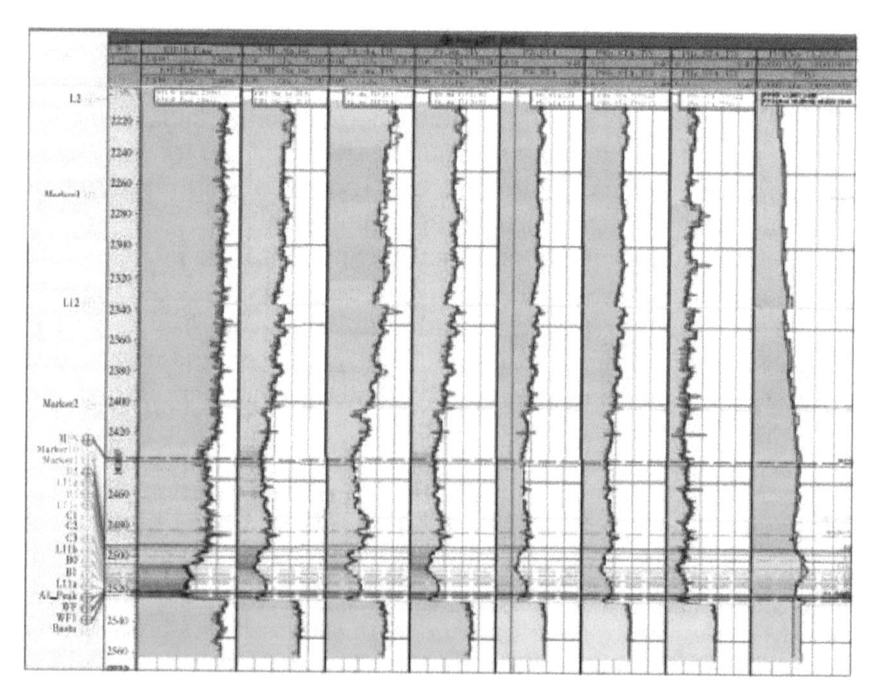

**Figure 7.39.** Comparison of 1D and 3D geomechanical parameters of well Ning-201 (from left to right: Depth, density, isotropic Young's modulus, transversal Young's modulus, vertical Young's modulus, isotropic Poisson's ratio, transversal Poisson's ratio, vertical Poisson's ratio, and pore pressure).

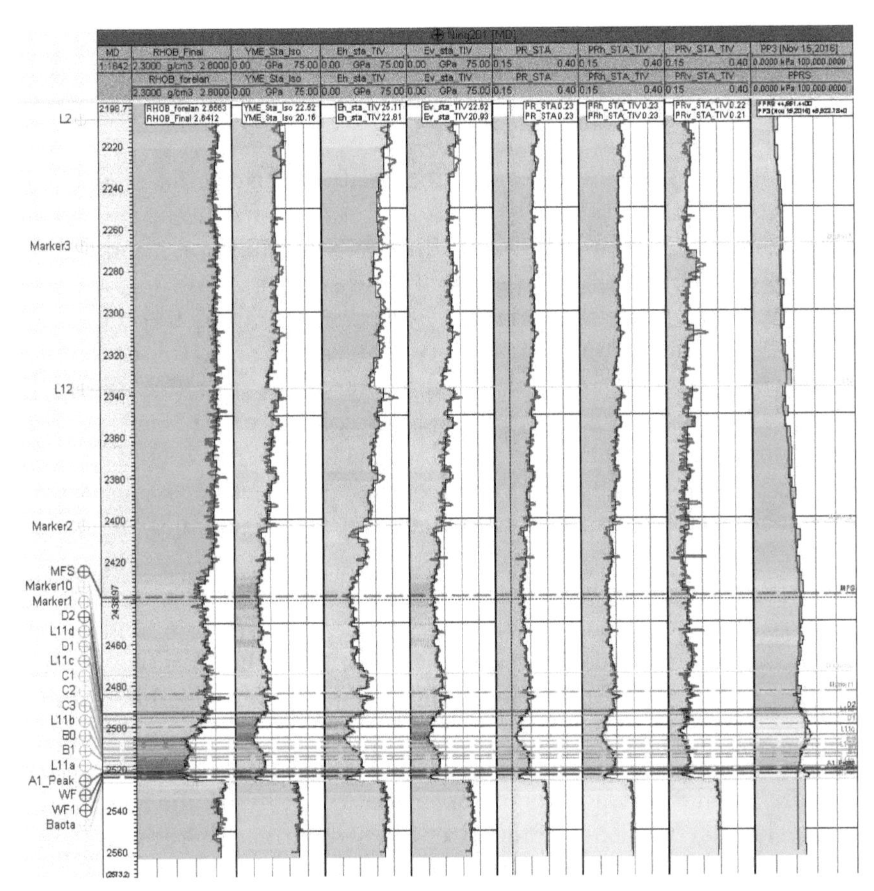


**Figure 7.40.** Comparison of 1D and 3D geomechanical parameters of well H9-2 (from left to right: depth, density, transversal Young's modulus, vertical Young's modulus, transversal Poisson's ratio, vertical Poisson's ratio, and pore pressure).

H9-2 (red curves represent the 1D model, and color-filled parts represent the 3D model). Since there was no Thrubit logging in well H9-2, acoustic anisotropy data lacked and the isotropic calculation results were used for comparison with lateral mechanical parameters. According to these figures, the 1D and 3D mechanical parameters were well matched throughout the interval.

Based on this idea, an integrated 3D geoscientific model was built through the comprehensive application of multi-disciplinary achievements such as log interpretation and geophysical, geological, and geomechanical modeling. This model will provide a series of 3D parameters for the drilling and completion engineering and can be used to optimize the well deployment scheme, predict drilling risks, support fracturing optimization, and analyze fracturing performance parameters.

Drilling and completion operations disturb the original stress field of rocks, leading to the redistribution of the *in situ* stress and possibly causing rock failure and accordingly increasing operational risks such as borehole wall collapse, drill pipe sticking, fluid losses, fault activity, and casing deformation. A series of parameters denoting the drilling and completion quality were determined based on the geomechanical model results of the whole study area, including 3D cracking pressure, collapse pressure, pore pressure, brittleness index, and stress barriers. They are visually shown in Figure 7.41. They can be used to support drilling, completion, and fracturing operations. Engineering applications of these parameters are as follows:

(1) After fracturing pressure and collapse pressure were extracted along the designed well trajectories, these parameters can be used to form a drilling fluid window to predict the optimal drilling fluid density and possible drilling complex events and determine corresponding preventive measures in the drilling process.
(2) These parameters can be used to provide input parameters for the simulation and scheme optimization of hydraulic fracturing, including brittleness index, tensile strength, and the minimum horizontal principal stress.
(3) These parameters provide valuable data support for the optimization of well deployment schemes, well trajectories, and fracturing process.

### 7.2.5. *Design optimization of horizontal well geosteering and fracturing construction scheme*

#### 7.2.5.1. *Compartment optimization and trajectory design of horizontal wells*

There is consensus on the importance of maximizing the probability of penetration of high-quality shale intervals. Figure 7.42 shows the relationship between the probability of penetration and the

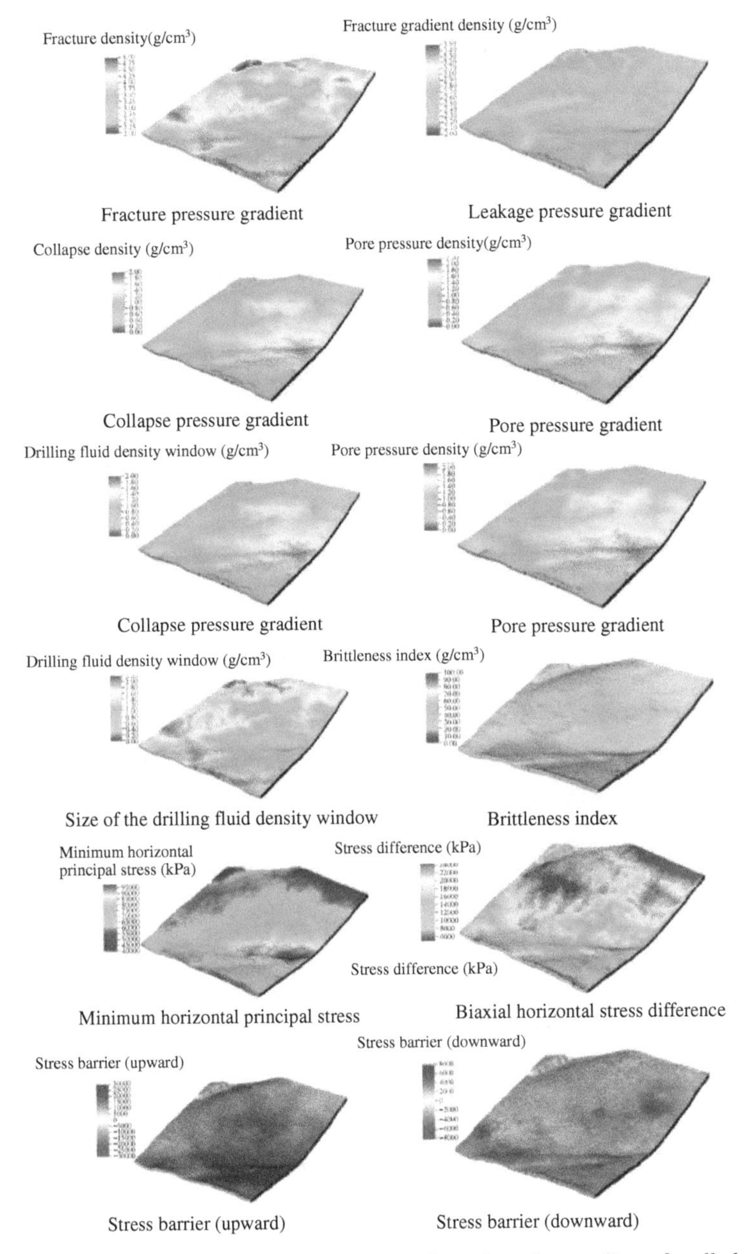


**Figure 7.41.** 3D diagrams of parameters denoting the quality of well drilling and completion in the whole study area.

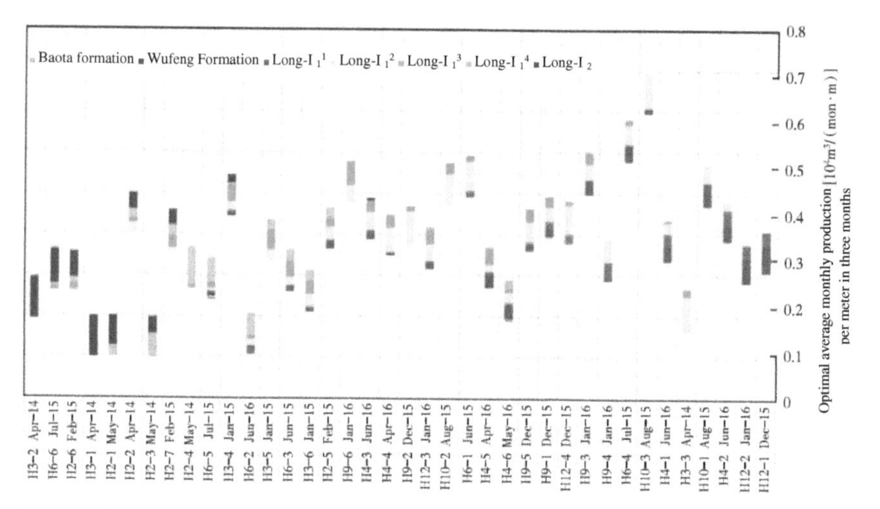


**Figure 7.42.** Relationship between the probability of penetration of various sublayers and the productivity (wells are ordered in decreasing average height from the bottom of the Wufeng Formation from left to right).

productivity of various sublayers at various wells in the Ning-201 gas field. The individual wells are ordered according to decreasing average height above the bottom of the Wufeng Formation from left to right. Each well is represented by a small colored column, in which different colors represent the different sublayers penetrated by the well and the proportions of different colors represent the probability of penetration of each sublayer. To a certain extent, the figure reflects the process for optimizing the location of horizontal well compartments in the Changning shale gas demonstration area. According to this figure, the productivity index tends to be higher in the horizontal sections closer to the bottom of the Wufeng Formation overall. Moreover, when horizontal sections are close to the bottom of the Wufeng Formation, the probability of penetration of high-quality shale intervals increases from the Wufeng Formation to the Long-$I_1$ Submember, with the productivity index also tending to increase accordingly. However, when the horizontal sections are mainly located in the high-quality shale interval from the Wufeng Formation to the Long-$I_1$ Submember, there is no significant correlation between the probability of penetration of various sublayers and the productivity index.

A high-resolution geoscientific conceptual model was built based on the log interpretation results of well Ning-201, as shown in Figure 7.43. Meanwhile, using the integrated fracturing — numerical

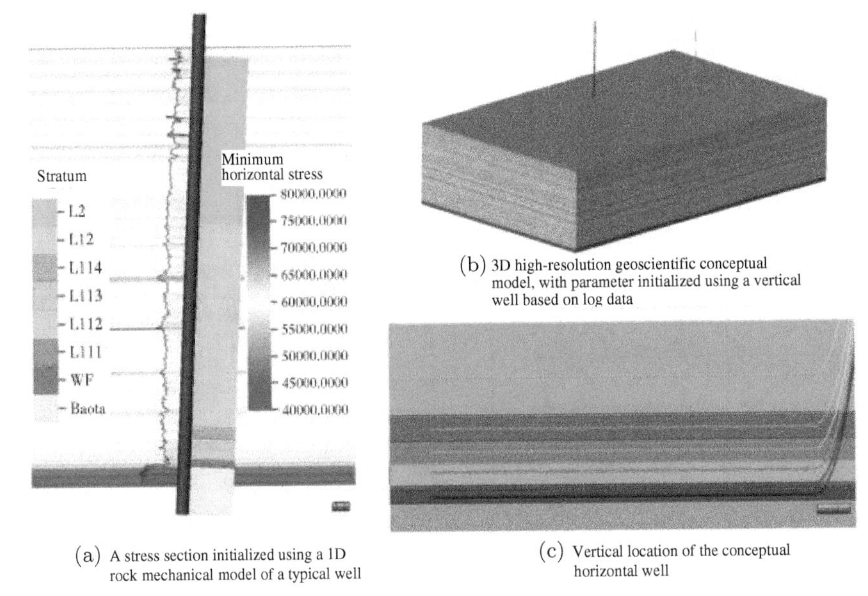


(a) A stress section initialized using a 1D rock mechanical model of a typical well

(b) 3D high-resolution geoscientific conceptual model, with parameter initialized using a vertical well based on log data

(c) Vertical location of the conceptual horizontal well

**Figure 7.43.** A high-resolution geoscientific conceptual model based on the well Ning-201.

simulation technology, this study carried out theoretical research on the effects of the compartment positions of horizontal wells on the hydraulic fracture network and productivity. The virtual horizontal wells in the model were used to completely penetrate the Long-$I_2$ Submember, the Long-$I_1^4$, Long-$I_1^3$, Long-$I_1^2$, and Long-$I_1^1$ Sublayers, and the sublayers of the lower submember of the Wufeng Formation. The horizontal interval length and the parameters for fracturing-related grades and clusters of the horizontal wells were set at the standard values currently used for the Ning-201 gas field. Hydraulic fracture parameters were controlled using values obtained from the fitting of microseismic data correction, fracturing construction response, and production history. The fracturing construction parameters included the pumping procedure, for which the current standard design scheme of the area of well Ning-201 was adopted. After the hydraulic fracture model was obtained, its numerical simulation model was built using an unstructured grid method. The INTERSECT simulator was used for productivity prediction, and the same pressure control production regime was used for all simulation models.

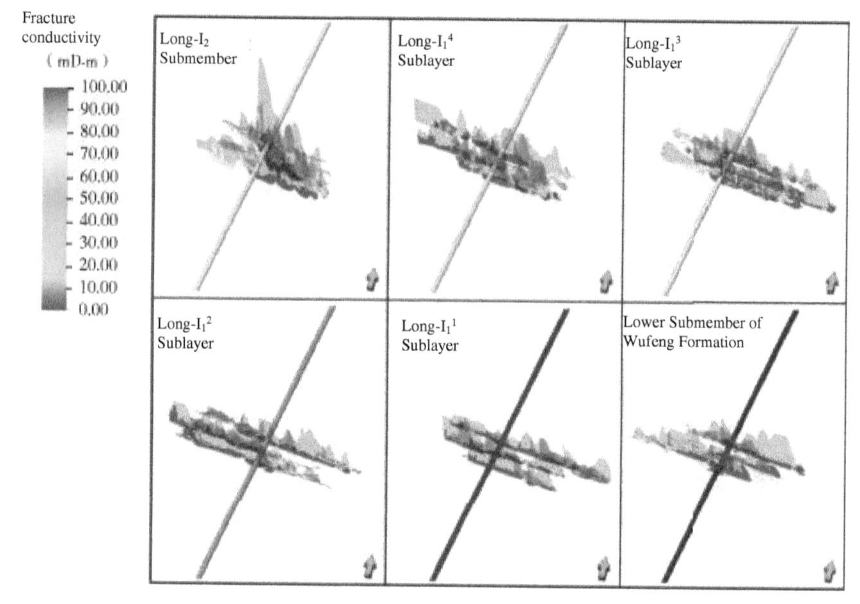


**Figure 7.44.** Geometric morphology of hydraulic fractures of different sublayers.

Through the comparative analysis of the morphology, effective propping volume, and proppant distribution of the hydraulic fracture network (Figure 7.44), it can be concluded that when all horizontal well sections of horizontal wells are located in the middle of the Long-$I_2$ Submember, fractures mainly extend upward to the shale intervals with poor quality due to the small stress of the upper part. Therefore, the high-quality shale intervals cannot be fully stimulated in this case. When all the horizontal well sections lie in the central part of the Long-$I_1^1$ Sublayer, the optimal stimulation performance can be theoretically obtained.

### 7.2.5.2. *Geosteering and trajectory adjustment of horizontal wells*

A structural model is an important tool for determining the geosteering direction of horizontal wells. However, due to the limitations of the uncertainties of seismic data caused by the complex terrains of mountainous areas, it is difficult to build an accurate structural model initially. Therefore, the structural model has undergone several update iterations, thereby continuously improving its structural accuracy. Practice shows that the structural model yields single-well depth errors during well drilling and completion of less than 15 m after continuous optimization and updating.

This study developed standard maps for pre-drilling design. They were derived from seismic interpretation or seismic data and can be used as a reference during well deployment. These standard maps mainly include:

(1) **Structural and depth maps:** They are used to determine the location of faults and the elevation and burial depth of targets;
(2) **Ant-tracking maps:** They are used to determine the locations of faults and fracture zones, which should be avoided as much as possible during well design;
(3) **Dip angle maps:** They are used to determine the dip angles of strata of the horizontal well sections and further calculate the build-up rate to provide a reference for the selection of geosteering instruments (e.g., determining steering tools according to the changes in the dip angle of horizontal well sections);
(4) **Structural curvature maps and microstructural maps:** They are used to show the microstructural characteristics, using which positive or negative microstructures can be determined. These maps, combined with the dip angles of strata, can be used to formulate the geosteering strategies of horizontal well sections.

The application of the seismic data-based integrated geoscientific model in the geosteering of horizontal well H25-10 is introduced as follows. Well H25-10 is adjacent to well H25-9, after the completion of which the structural model was updated. As revealed by well H25-9, a fault zone with a throw of approximately 30 m was encountered on well landing (Figure 7.45), following which the structures significantly changed. The dip structures along the horizontal section of well H25-9 are relatively consistent, with a dip angle of approximately 10°.

The design review in January 2017 indicated that well H25-10 was predicted to have similar structures to well H25-9. In other words, its portion from the build-up section to the landing location is affected by faults and shows strong anomalies of ant-tracking volumes, and its horizontal section inclines upward at an angle of 98°–102°. Based

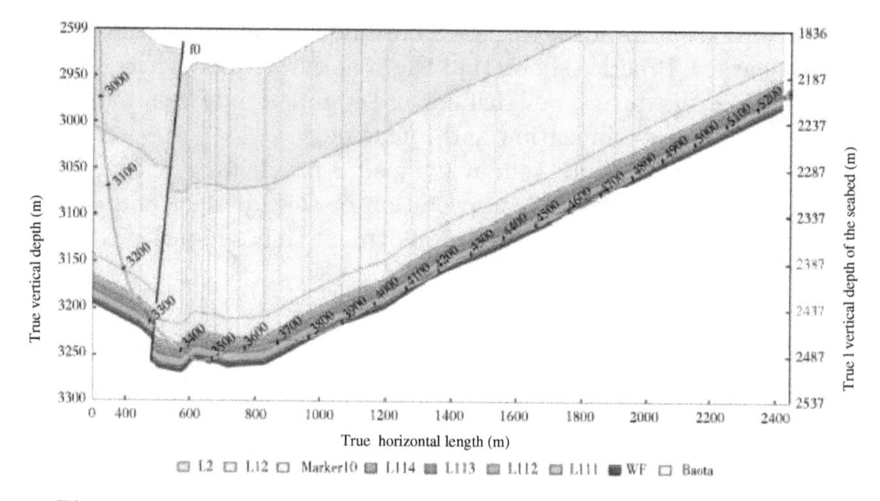


Figure 7.45. Model section along the horizontal portion of well H25-9.

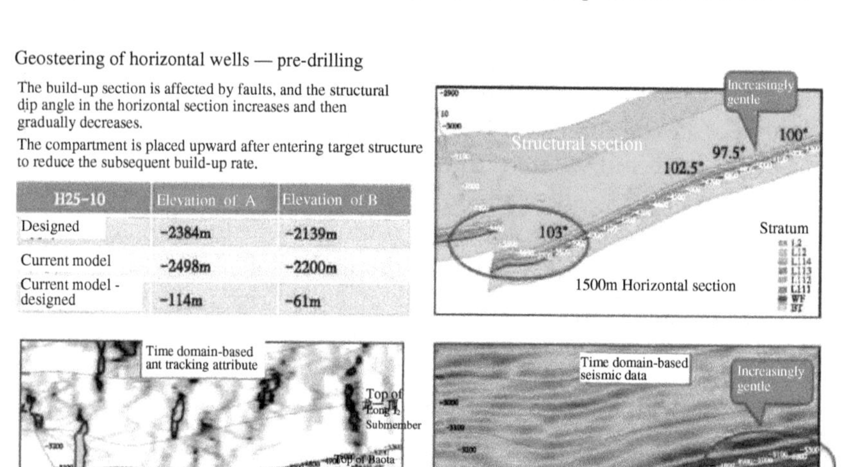


Figure 7.46. Pre-drilling model of the well H25-10.

on these findings, the depth of the targets was updated according to the structural model (Figure 7.46).

On May 3, 2017, the structural model was updated and followed up before well H25-10 entered the target, and the well trajectory in the updated model was predicted. The top of the Middle Ordovician

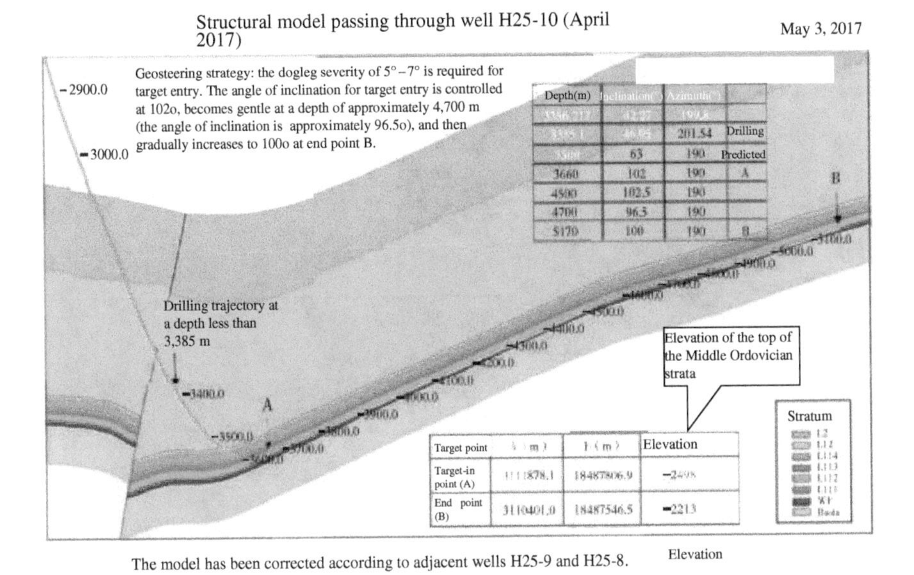


**Figure 7.47.** Section of the structural model while drilling of well H25-10.

strata at target-in point A and end point B was predicted to have an elevation of −2,498 m and −2,213 m, respectively (Figure 7.47). Based on this model, it was calculated that a dogleg severity of 5° − 7° was required for target entry. The structural trend along the whole horizontal portion of the well was predicted by integrating the seismic section and the ant-tracking section (Figure 7.48).

The structural section on drilling completion (Figure 7.49) proves that encouraging geosteering performance of well H25-10 has been achieved, with the Long-$I_1^1$ Sublayer and the top of the Wufeng Formation being mainly encountered. The elevation of the top of the Middle Ordovician before landing at points A and B was predicted to be −2,498 m and −2,213 m, respectively. It was −2,496 m and −2,226 m, respectively, according to drilling results, with errors of 2 m and 13 m, respectively. Meanwhile, it is noteworthy that the seismic section shows positive microstructures developing in the central portion of well H25-10, but the drilling model shows a monocline with a dip angle of approximately 10° in this portion. The reason for the deviation in the seismic section may be related to the large faults and the high dip angles of strata near the well.

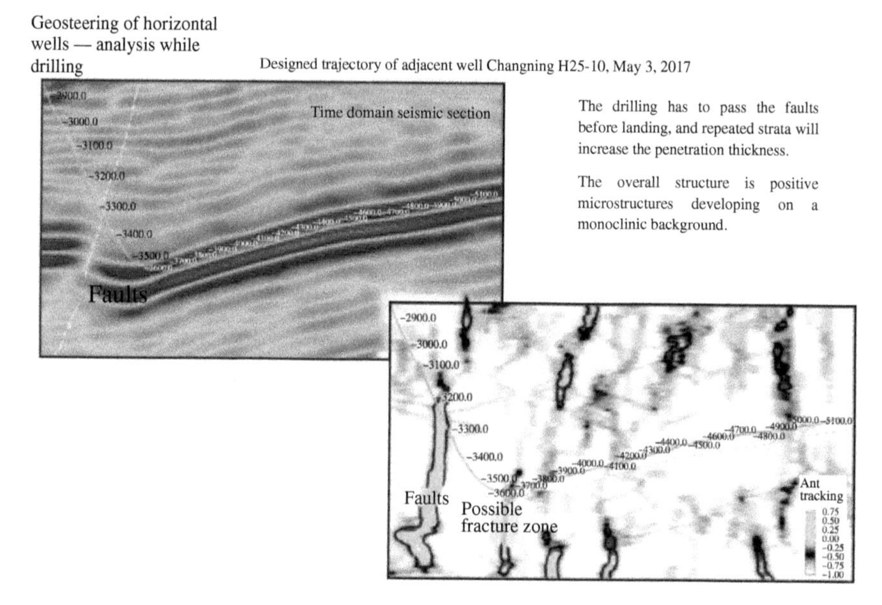


**Figure 7.48.** Section of seismic attributes while drilling of well H25-10.

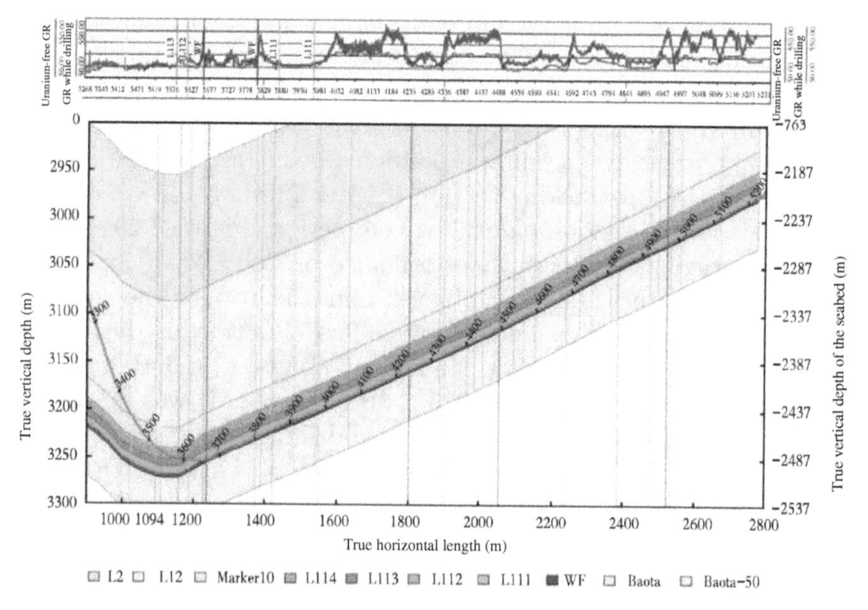


**Figure 7.49.** Trajectory section of the drilled well H25-10.

#### 7.2.5.3. *Optimization of the fracturing construction scheme and evaluation of the fracturing performance*

The fracturing construction process is related to factors such as *in situ* stress, rock brittleness, rock strength, and pore pressure. After a 3D geomechanical model is built, it is necessary to analyze and explore the correlation between the mechanical parameters and fracturing construction parameters. The purpose is to summarize the fracturing construction experience and to provide an engineering-related reference for subsequent fracturing constructions.

As indicated by the study in this section, the parameters related to hydraulic fracturing (including construction pressure, instantaneous shut-in pressure (ISIP), and pumped sand amount) are affected by the density of natural fractures and *in situ* stress, while the *in situ* stress is affected by multiple factors such as burial depth, pore pressure, and fractures. These main control factors can be fully considered by utilizing the integrated geological–engineering modeling method and the geomechanical simulation results, thus optimizing the fracturing construction design.

Figure 7.50 shows the correlations between the pumped sand amount and parameters such as stress, pore pressure, and ant-tracking attributes at wellpad H4. The minimum horizontal principal stress around the wellpad is high in the southeast and low in the northwest, which is consistent with the stress distribution pattern of the whole study area. Under the influence of pore pressure, the biaxial stress difference tends to increase from SE to NW, but the differential stress variation around the wellpad is not significant. The average pumped sand amount differed slightly at a certain stage between the wells of wellpad H4 but differed greatly between different stages. There is a close correlation between the fracturing stage with a low pumped sand amount and the ant-tracking attributes. This result indicates that the density of natural fractures may lead to loss of fracturing fluids, increased risks of sand plugging, and increased construction pressure.

The strength of the ant-tracking attributes is insufficient to fully explain the effects of natural fractures. Figure 7.51 shows the correlation between the fracturing construction parameters of the H4 wellpad and natural fractures. Figure 7.51(a) shows the comparison between pumped sand amount and fracture stability. The red

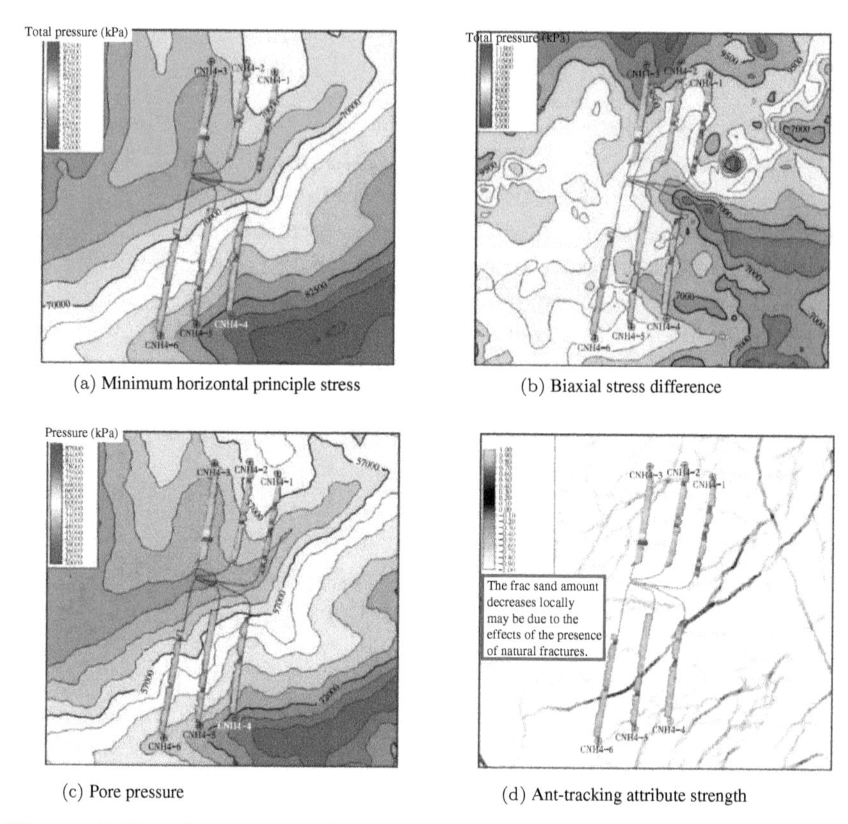


(a) Minimum horizontal principle stress

(b) Biaxial stress difference

(c) Pore pressure

(d) Ant-tracking attribute strength

**Figure 7.50.** Comparison of sand amounts for H4 wellpad with stress, pore pressure, and ant-tracking attribute.

circles indicate fracturing stages at which fractures are prone to dislocation and sand plugging and the pumped sand amount is lower than that of the adjacent fracturing stages. The blue circle indicates that the developed fractures have high stability and are not prone to dislocation. In this case, the fracturing fluids are not prone to lose, and the pumped sand amount is not significantly affected. Figure 7.51(b) shows the ISIP and the minimum horizontal principal stress of six wells of the H4 wellpad, with filled colors representing the ant-tracking attribute strength. According to this figure, there is a positive correlation between the ISIP and the minimum horizontal principal stress. In other words, higher minimum horizontal principal stress is associated with greater ISIP. The points

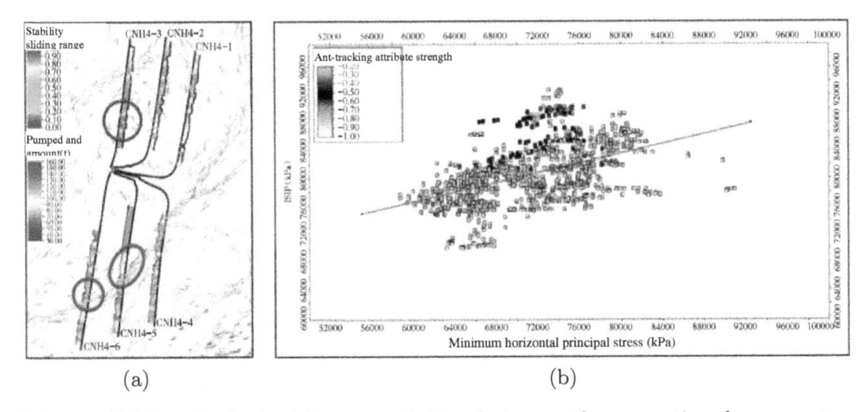


(a) (b)

**Figure 7.51.** Analysis of the correlation between the operational parameters of the H4 wellpad and natural fractures.

with high ant-tracking attribute strength are located above the fitted line, indicating that the natural fractures have high ISIP. The ISIP reflects the pressure required to crack the fracture plane and equals the sum of the normal stress on the fracture plane and net pressure. When hydraulic fractures connect to natural fractures, the normal stress of the natural fracture plane should be overcome to crack the fracture plane. The ISIP mainly reflects the complex stress state near the wellbore rather than the original stress field. Therefore, natural fractures with high density are associated with higher ISIP. The difference between the fracturing construction pressure and ISIP lies in the friction resistance. Therefore, the fracturing construction pressure has similar controlling factors to ISIP and thus, is not discussed here.

## Bibliography

Bowker KA. Barnett Shale gas production, Fort Worth Basin: Issues and discussion. *AAPG Bulletin*, 2007, 91 (4): 523–533.

Chen GS, Wu JF, Liu Y, *et al.* Geology-engineering integration key technologies for ten billion cubic meters of shale gas productivity construction in the Southern Sichuan Basin. *Natural Gas Industry*, 2021, 41 (1): 72–82.

Curtis JB. Fractured shale gas systems. *AAPG Bulletin*, 2002, 86 (11): 1921–1938.

Gale JFW, Holder J. Natural fractures in some U. S. shales and their importance for gas production. In Vining BA, Pickering SC (eds.), *Petroleum Geology: From Mature Basins to New Frontiers*. London: The Geological Society, 2010, pp. 1131–1140.

Guo TL, Zeng P. The structural and preservation conditions for shale gas enrichment and high productivity in the Wufeng-Longmaxi Formation, Southeastern Sichuan Basin. *Energy Exploration & Exploitation*, 2015, 33 (3): 259–276.

Guo TL. Evaluation of highly thermally mature shale-gas reservoirs in complex structural parts of the Sichuan Basin. *Journal of Earth Science*, 2013, 24 (6): 863–873.

Guo TL. Key geological issues and main controls on accumulation and enrichment of Chinese shale gas. *Petroleum Exploration and Development*, 2016, 43 (3): 317–326.

Guo TL. The Fuling shale gas field: A highly productive Silurian gas shale with high thermal maturity and complex evolution history, southeastern Sichuan Basin, China. *Interpretation*, 2015, 3 (2): 1–10.

Hammes U, Hamlin HS, Ewing TE. Geologic analysis of the Upper Jurassic Haynesville shale in east Texas and west Louisiana. *AAPG Bulletin*, 2011, 95 (10): 1643–1666.

Hu WR. Geology-engineering integration-a necessary way to realize profitable exploration and development of complex reservoirs. *China Petroleum Exploration*, 2017, 22 (1): 1–5.

Huang HY, Fan Y, Zeng B, *et al.* Geology-engineering integration of platform well in Changning Block. 2020, 20 (1): 175–182.

Hughes JD. Energy: A reality check on the shale revolution. *Nature*, 2013, 494 (7437): 307–308.

Jarvie DM. Shale resource systems for oil and gas: Part 1: Shale-gas resource systems. In Breyer JA (ed.), *Shale Reservoirs: Giant Resources for the 21st Century: AAPG Memoir 97*. Tulsa: AAPG, 2012, pp. 69–87.

Liang X, Wang GC, Jiao YJ, *et al.* Zhaotong national shale gas pilot block case study: Integrated geological modeling and its application for sweet spot prediction in shale gas reservoir. In *SPG/SEG Beijing 2016 International Geophysical Conference*, Beijing, China, April 2016.

Liang X, Xu ZY, Zhang JH, *et al.* Key efficient exploration and development technologies of shallow shale gas: A case study of Taiyang anticline area of Zhaotong National Shale gas demonstration zone. *Acta Petrolei Sinica*, 2020, 41 (9): 1033–1048.

Lin XJ, Hu SY, Cheng KM. Suggestions from the development of fractured shale gas in North America. *Petroleum Exploration and Development*, 2007, 34 (4): 392–400.

Liu H, Li GX, Yao ZX, *et al.* "Point-line-area" methodology of shale oil exploration and development. *Oil Forum*, 2020, 39 (2): 1–5.

Liu N, Wang G. Shale gas sweet spot identification and precise geo-steering drilling in Weiyuan Block of Sichuan Basin, SW China. *Petroleum Exploration & Development*, 2016, 43 (6): 1067–1075.

Liu NZ, Wang GY, Xiong XL, *et al.* Practice and prospect of geology-engineering integration technology in the efficient development of shale gas in Weiyuan block. *China Petroleum Exploration*, 2018, 115 (2): 63–72.

Liu QY, Zhu HY, Chen PJ. Research progress and direction of geology-engineering integrated drilling technology: A case study on the deep shale gas reservoirs in the Sichuan Basin. *Natural Gas Industry*, 2021, 41 (1): 178–188.

Liu SG, Ran B, Guo TL, *et al.* *Lower Palaeozoic Organic-Matter-Rich Black Shale in the Sichuan Basin and Its Periphery: From Oil-Prone Source Rock to Gas-Producing Shale Reservoir.* Beijing: Science Press, 2014.

Pan S, Zou C, Yang Z, *et al.* Methods for shale gas play assessment: A comparison between Silurian Longmaxi shale and Mississippian Barnett shale. *Journal of Earth Science*, 2015, 26: 285–294.

Peggy W. Big sandy. *Oil and Gas Investor*, 2005, 8: 73–75.

Rickman R, Mullen M, Petre E, *et al.* A practical use of shale petrophysics for stimulation design optimization: All shale plays are not clones of the Barnett shale. In *The SPE Annual Technical Conference and Exhibition*, Dallas, Texas, USA, September 2018.

Shu HL, Wang LZ, Yin KG, *et al.* Geological modeling of shale gas reservoir during the implementation process of geology-engineering integration. *Geological Modeling of Shale Gas Reservoir During the Implementation Process of Geology-Engineering Integration*, 2020, 25 (2): 84–95.

Su WB, He LQ, Wang YB, *et al.* Biostratigraphy and geography of the Ordovician-Silurian Longmaxi black shales in South China. *Science in China: Earth Sciences*, 2002, 32 (3): 207–219.

Wang XZ, Zhang JC, Cao JZ, *et al.* A preliminary discussion on evaluation of continental shale gas resources: A case study of Chang 7 of Mesozoic Yanchang Formation in Zhiluo-Xiasiwan area of Yanchang. *Earth Science Frontiers*, 2012, 19 (2): 192–197.

Wang ZP, Zhang JC, Sun R, *et al.* The gas-bearing characteristics analysis of the Longtan Formation transitional shale in Well Xiye 1. *Earth Science Frontiers*, 2015, 22 (2): 243–250.

Wu Q, Liang X, Xian CG, *et al.* Geoscience-to-production integration ensures effective and efficient South China marine shale gas development. *China Petroleum Exploration*, 2015, 20 (4): 1–23.

Xie J, Xian CG, Wu JF, *et al.* Optimal key elements of geoengineering integration in Changning National Shale Gas Demonstration Zone. *China Petroleum Exploration*, 2019, 24 (2): 174–185.

Xie J. Practices and achievements of the Changning–Weiyuan shale gas national demonstration project construction. *Natural Gas Industry*, 2018, 38 (2): 1–7.

Zagorski WA, Wrightstone GR, Bowman DC. The Appalachian Basin Marcellus gas play: Its history of development, geologic controls on production, and future potential as a world-class reservoir. In Breyerj A (ed.), *Shale Reservoirs: Giant Resources for the 21st Century: AAPG Memoir 97*. Tulsa: AAPG, 2012, pp. 172–200.

Zha SG, Liu LP, Liao P, *et al.* Seismic geo-steering technology of horizontal well and its application in Fuling shale gas field. *Geophysical Prospecting for Petroleum*, 2018, 57 (3): 52–60.

Zhao JZ, Fang ZQ, Zhang J, *et al.* Evaluation of China shale gas from the exploration and development of North America shale gas. *Journal of Xi'an Shiyou University*, 2011, 26 (2): 1–7.

Zhou DS, Xu LF, Pan JP, *et al.* Prospect of shale gas exploration in the Upper Permian Longtan Formation in the Yangtze Massif. *Natural Gas Industry*, 2012, 23 (12): 6–10.

Zhu DX, Jiang LW, Niu WT, *et al.* Seismic and geological integration applied in the shale gas exploration. *Oil Geophysical Prospecting*, 2018, 53 (S1): 249–255, 17.

Zou CN, Yang Z, Zhu RK, *et al.* Progress in China's unconventional oil & gas exploration and development and theoretical technologies. *Acta Geologica Sinica*, 2015, 89 (6): 979–1007.

Zou CN, Zhang GS, Yang Z, *et al.* Geological concepts, characteristics, resource potential and key techniques of unconventional hydrocarbon: On unconventional petroleum geology. *Petroleum Exploration and Development*, 2013, 40 (4): 385–399.

https://doi.org/10.1142/9789811283185_bmatter

# Index

**L**



**M**



**N**



**P**



**Q**



**R**



**S**



**T**

Printed in the USA
CPSIA information can be obtained
at www.ICGtesting.com
LVHW051417250724
786421LV00002B/28

9 789811 283178